THE SCIENCE AND ETHICS OF GENETICALLY ENGINEERED HUMAN DNA

HEARING

BEFORE THE

SUBCOMMITTEE ON RESEARCH & TECHNOLOGY

COMMITTEE ON SCIENCE, SPACE, AND TECHNOLOGY

HOUSE OF REPRESENTATIVES

ONE HUNDRED FOURTEENTH CONGRESS

FIRST SESSION

JUNE 16, 2015

Serial No. 114–24

Printed for the use of the Committee on Science, Space, and Technology

Available via the World Wide Web: http://science.house.gov

U.S. GOVERNMENT PUBLISHING OFFICE

97–564PDF   WASHINGTON : 2016

For sale by the Superintendent of Documents, U.S. Government Publishing Office
Internet: bookstore.gpo.gov   Phone: toll free (866) 512–1800; DC area (202) 512–1800
Fax: (202) 512–2104   Mail: Stop IDCC, Washington, DC 20402–0001

COMMITTEE ON SCIENCE, SPACE, AND TECHNOLOGY

HON. LAMAR S. SMITH, Texas, *Chair*

FRANK D. LUCAS, Oklahoma
F. JAMES SENSENBRENNER, JR., Wisconsin
DANA ROHRABACHER, California
RANDY NEUGEBAUER, Texas
MICHAEL T. McCAUL, Texas
MO BROOKS, Alabama
RANDY HULTGREN, Illinois
BILL POSEY, Florida
THOMAS MASSIE, Kentucky
JIM BRIDENSTINE, Oklahoma
RANDY K. WEBER, Texas
BILL JOHNSON, Ohio
JOHN R. MOOLENAAR, Michigan
STEVE KNIGHT, California
BRIAN BABIN, Texas
BRUCE WESTERMAN, Arkansas
BARBARA COMSTOCK, Virginia
DAN NEWHOUSE, Washington
GARY PALMER, Alabama
BARRY LOUDERMILK, Georgia
RALPH LEE ABRAHAM, Louisiana

EDDIE BERNICE JOHNSON, Texas
ZOE LOFGREN, California
DANIEL LIPINSKI, Illinois
DONNA F. EDWARDS, Maryland
SUZANNE BONAMICI, Oregon
ERIC SWALWELL, California
ALAN GRAYSON, Florida
AMI BERA, California
ELIZABETH H. ESTY, Connecticut
MARC A. VEASEY, TEXAS
KATHERINE M. CLARK, Massachusetts
DON S. BEYER, JR., Virginia
ED PERLMUTTER, Colorado
PAUL TONKO, New York
MARK TAKANO, California
BILL FOSTER, Illinois

SUBCOMMITTEE ON RESEARCH AND TECHNOLOGY

HON. BARBARA COMSTOCK, Virginia, *Chair*

FRANK D. LUCAS, Oklahoma
MICHAEL T. MCCAUL, Texas
RANDY HULTGREN, Illinois
JOHN R. MOOLENAAR, Michigan
BRUCE WESTERMAN, Arkansas
DAN NEWHOUSE, Washington
GARY PALMER, Alabama
RALPH LEE ABRAHAM, Louisiana
LAMAR S. SMITH, Texas

DANIEL LIPINSKI, Illinois
ELIZABETH H. ESTY, Connecticut
KATHERINE M. CLARK, Massachusetts
PAUL TONKO, New York
SUZANNE BONAMICI, Oregon
ERIC SWALWELL, California
EDDIE BERNICE JOHNSON, Texas

CONTENTS

June 16, 2015

	Page
Witness List	2
Hearing Charter	3

Opening Statements

Statement by Representative Barbara Comstock, Chairwoman, Subcommittee on Research, Committee on Science, Space, and Technology, U.S. House of Representatives 5
    Written Statement 5

Statement by Representative Daniel Lipinski, Ranking Minority Member, Subcommittee on Research, Committee on Science, Space, and Technology, U.S. House of Representatives 6
    Written Statement 7

Statement by Representative Lamar S. Smith, Chairman, Committee on Science, Space, and Technology, U.S. House of Representatives 8
    Written Statement 9

Witnesses:

Dr. Victor J. Dzau, President, Institute of Medicine, the National Academy of Sciences
    Oral Statement 10
    Written Statement 13

Dr. Jennifer Doudna, Professor of Biochemistry and Molecular Biology, University of California,
    Oral Statement 18
    Written Statement 20

Dr. Elizabeth McNally, Professor of Genetic Medicine, Professor in Medicine-Cardiology and Biochemistry and Molecular Genetics; Director, Center for Genetic Medicine, Northwestern University
    Oral Statement 23
    Written Statement 25

Dr. Jeffrey Kahn, Professor of Bioethics and Public Policy; Deputy Director for Policy and Administration, Berman Institute of Bioethics, Johns Hopkins University
    Oral Statement 32
    Written Statement 34

Discussion 40

Appendix I: Answers to Post-Hearing Questions

Dr. Elizabeth McNally, Professor of Genetic Medicine, Professor in Medicine-Cardiology and Biochemistry and Molecular Genetics; Director, Center for Genetic Medicine, Northwestern University 64

Appendix II: Additional Material for the Record

Statement by Representative Eddie Bernice Johnson, Ranking Member, Committee on Science, Space, and Technology, U.S. House of Representatives 66

THE SCIENCE AND ETHICS OF GENETICALLY ENGINEERED HUMAN DNA

TUESDAY, JUNE 16, 2015

House of Representatives,
Subcommittee on Research and Technology
Committee on Science, Space, and Technology,
Washington, D.C.

The Subcommittee met, pursuant to call, at 2:16 p.m., in Room 2318 of the Rayburn House Office Building, Hon. Barbara Comstock [Chairwoman of the Subcommittee] presiding.

LAMAR S. SMITH, Texas
CHAIRMAN

EDDIE BERNICE JOHNSON, Texas
RANKING MEMBER

Congress of the United States
House of Representatives

COMMITTEE ON SCIENCE, SPACE, AND TECHNOLOGY

2321 RAYBURN HOUSE OFFICE BUILDING

WASHINGTON, DC 20515-6301

(202) 225-6371

www.science.house.gov

Subcommittee on Research and Technology

The Science and Ethics of Genetically Engineered Human DNA

Tuesday, June 16, 2015
2:00 p.m. to 4:00 p.m.
2318 Rayburn House Office Building

Witnesses

Dr. Victor J. Dzau, *President, Institute of Medicine, the National Academy of Sciences*

Dr. Jennifer Doudna, *Professor of Biochemistry and Molecular Biology, University of California, Berkeley*

Dr. Elizabeth McNally, *Professor of Genetic Medicine, Professor in Medicine-Cardiology and Biochemistry and Molecular Genetics; Director, Center for Genetic Medicine, Northwestern University*

Dr. Jeffrey Kahn, *Professor of Bioethics and Public Policy; Deputy Director for Policy and Administration, Berman Institute of Bioethics, Johns Hopkins University*

U.S. HOUSE OF REPRESENTATIVES
COMMITTEE ON SCIENCE, SPACE, AND TECHNOLOGY
SUBCOMMITTEE ON RESEARCH AND TECHNOLOGY

HEARING CHARTER

The Science and Ethics of Genetically Engineered Human DNA

Tuesday, June 16, 2015
2:00 p.m. – 4:00 p.m.
2318 Rayburn House Office Building

Purpose

On Tuesday, June 16, 2015, the Research & Technology Subcommittee will hold a hearing titled *The Science and Ethics of Genetically Engineered Human DNA*. Human gene editing has been a major topic of news and editorials in recent months in both the popular and scientific press.[1] The purpose of the hearing is to review the science behind new gene editing technologies, examine the ethical implications and risks, discuss the promise or potential applications for this new research, and explore how to build a responsible framework for utilizing gene editing technologies. Witnesses will also discuss how the United States can provide scientific and ethical leadership in this arena.

Witness List

Dr. Victor Dzau, President, Institute of Medicine
Dr. Jennifer Doudna, Professor of Biochemistry and Molecular Biology, University of California, Berkeley
Dr. Elizabeth McNally, Director, Center for Genetic Medicine, Northwestern University
Dr. Jeffrey Kahn, Professor of Bioethics and Public Policy and Deputy Director for Policy and Administration of the Berman Institute of Bioethics, Johns Hopkins University

Background

Genome-editing tools that allow a gene to be deleted, inserted, or replaced by a different piece of DNA are becoming more cost-effective and simpler to execute. New gene-editing techniques that can repair or enhance a human gene, are also now capable of altering the human germline – the cells that last for the life of the individual and are passed on to future generations.[2]

---

[1] "CRISPR: The Good, the Bad, the Unknown," *Nature* Special Archive. Available at: http://www.nature.com/news/crispr-1.17547
"Scientists are Growing Anxious about Genome-Editing Tools," *Washington Post*, May 18, 2015. Available at: http://www.washingtonpost.com/national/health-science/scientists-are-growing-anxious-about-genome-editing-tools/2015/05/18/0a4db63c-ef4e-11e4-8abc-d6aa3bad79dd_story.html
[2] "A Powerful New Way to Edit DNA," *New York Times*, March 3, 2014. Available at: http://www.nytimes.com/2014/03/04/health/a-powerful-new-way-to-edit-dna.html

Last April, it was reported that a team of Chinese researchers attempted to edit the genome of human embryos for the first time.[3] The team used a new gene-editing technology called CRISPR/Cas9 in an attempt to replace a gene in 86 non-viable human embryos. The technique was successful in only a small fraction of the embryos and caused other unintended genetic mutations.[4] In the wake of the Chinese team publishing their study, the National Institutes of Health issued a statement that in the United States there are "existing legislative and regulatory prohibitions against this kind of work."[5]

Many in the scientific community, including prominent scientists and inventors of gene-editing technologies, have called for a worldwide moratorium on such altering of human DNA.[6] Although these technologies may promise new treatments for inherited genetic diseases such as cystic fibrosis, sickle cell disease and hemophilia, there is also concern that they could be abused to create "designer babies" or alter heritable human DNA in unexpected or dangerous ways.[7]

Precedents for this type of moratorium exist. In 1975, the scientific community agreed on a worldwide suspension on experimenting with new recombinant DNA techniques to manipulate genes, while a safety and ethical framework was developed. The safe and ethical use of this technology later became the underpinning of the biotechnology industry, creating new medical treatments, agriculture products, and biofuels.[8]

Last month, the National Academy of Science (NAS) and Institute of Medicine (IOM, to be renamed the National Academy of Medicine effective July 1st) announced that they were launching a major "Initiative on Human Gene Editing."[9] The organizations are proposing a comprehensive study of the scientific underpinnings and clinical, ethical, legal and social implications of human gene editing. NAS and IOM will also recommend an international symposium of stakeholders to discuss setting guidelines for such genome-editing research.

---

[3] "Chinese Scientists Edit Genes of Human Embryos, Raising Concerns," *New York Times*, April 23, 2015. Available at: http://www.nytimes.com/2015/04/24/health/chinese-scientists-edit-genes-of-human-embryos-raising-concerns.html
[4] "Chinese Scientists Genetically Modify Human Embryos," *Nature*, April 22, 2015. Available at: http://www.nature.com/news/chinese-scientists-genetically-modify-human-embryos-1.17378
[5] Statement from Dr. Francis Collins, Director, National Institutes of Health, April 29, 2015. Available at: http://www.nih.gov/about/director/04292015_statement_gene_editing_technologies.htm
[6] "A Prudent Path Forward for Genomic Engineering and Germline Gene Modification," *Science Magazine*, April 3, 2015. Available at: http://www.sciencemag.org/content/348/6230/36
"Let's Hit 'Pause' Before Altering Humankind," *Wall Street Journal*, April 8, 2015. Available at: http://www.wsj.com/articles/lets-hit-pause-before-altering-humankind-1428536400
International Society for Stem Cell Research Statement in support of moratorium. Available at: http://www.isscr.org/home/about-us/news-press-releases/2015/2015/03/19/statement-on-human-germline-genome-modification
[7] "Engineering the Perfect Baby," *MIT Technology Review*, March 5, 2015. Available at: http://www.technologyreview.com/featuredstory/535661/engineering-the-perfect-baby/
[8] NIH Human Genome Project History. Available at: http://www.genome.gov/25520302
[9] National Academy of Sciences and National Academy of Medicine Announce Initiative on Human Gene Editing. Available at: http://www.iom.edu/Global/News%20Announcements/NAS-NAM-Human-Gene-Editing.aspx

Chairwoman COMSTOCK. The Subcommittee on Research and Technology will come to order. Without objection, the Chair is authorized to declare recesses of the Subcommittee at any time.

And without objection, the gentleman from California, Mr. Sherman, is authorized to participate in today's hearing.

Good afternoon, and welcome to this hearing entitled "The Science and Ethics of Genetically Engineered Human DNA."

And I believe we also would like to welcome Representative Abraham to his first Science Committee hearing. Dr. Abraham, we are happy to have you join the Research and Technology Subcommittee and we look forward to having the benefit of your expertise.

Now, in front of you are packets containing the written testimonies, biographies, and truth-in-testimony disclosures for today's witnesses.

I now recognize myself for an opening statement.

Biotechnology—the engineering of genetic material in living beings and plants—has transformed modern medicine and agriculture. Rapid advances in biotech research have brought great opportunities for new medical treatments and products, and simultaneously have also raised questions about possible ethical implications and safety issues.

Today, we are here to discuss the science and ethics of the most recent and eye-opening development in biotechnology: human genome-editing. This research has been a major topic of news and editorials in recent months. New tools that allow a gene to be deleted, inserted, or replaced by a different piece of DNA are becoming more cost-effective and simpler to execute.

In April it was reported that for the first time a team of Chinese scientists had attempted to edit the genome of human embryos. The report raised concerns for many scientists and policymakers about the safety and ethics of using these new technologies on human DNA. Many prominent scientists have called for a better framework to be developed for responsible use of the technology.

I look forward to learning more from our witnesses today who will provide an overview of the science behind these new technologies, help us examine the implications and risks, and explore what the next steps should be for building the right kind of framework for utilizing the technology. They will also help us answer how the United States can be a leader and provide scientific and ethical leadership in this arena.

[The prepared statement of Chairwoman Comstock follows:]

PREPARED STATEMENT OF SUBCOMMITTEE
CHAIRWOMAN BARBARA COMSTOCK

Biotechnology—the engineering of genetic material in living beings and plants—has transformed modern medicine and agriculture.

Rapid advances in biotech research have brought great opportunities for new medical treatments and products, and simultaneously have also raised questions about possible ethical implications and safety issues.

Today, we are here to discuss the science and ethics of the most recent and eye-opening development in biotechnology: human genome-editing.

This research has been a major topic of news and editorials in recent months. New tools that allow a gene to be deleted, inserted, or replaced by a different piece of DNA are becoming more cost-effective and simpler to execute.

In April, it was reported that for the first time a team of Chinese scientists had attempted to edit the genome of human embryos. The report raised concerns for many scientists and policy makers about the safety and ethics of using these new technologies on human DNA.

Many prominent scientists have called for a better framework to be developed for responsible use of the technology.

I look forward to learning more from our witnesses today who will provide an overview of the science behind these new technologies, help us examine the ethical implications and risks, and explore what the next steps should be for building a responsible framework for utilizing the technology. They will also help us answer how the United States can provide scientific and ethical leadership in this arena.

Chairwoman COMSTOCK. So I now recognize the Ranking Member, the gentleman from Illinois, for his opening statement.

Mr. LIPINSKI. Thank you, Chairwoman Comstock, for holding this hearing on the science and ethics of new gene editing technologies.

I want to thank all the witnesses for being here today and look forward to your testimony.

Although we're talking about gene editing technologies that are very new, it's important to mention that humans have been altering the genomes of species through selective breeding for thousands of years. And since the 1970s, it has been possible to directly manipulate DNA, which led to a biotechnology revolution and significant economic growth.

Then we had the Human Genome Project to sequence the human genome, and it was coordinated by the Department of Energy and the National Institutes of Health. The full human genome was sequenced in 2003, opening up whole new possibilities for diagnosing and treating diseases. One such pathway led to the invention of the CRISPR technology.

Thanks to new gene editing technologies, which include CRISPR, we're able to add, remove, and replace DNA bases. They can be thought of as search-and-replace tools for DNA. They're incredibly powerful technologies that have the potential to transform the healthcare, energy, and agricultural sectors. Although new, these technologies were the outgrowth of decades of fundamental research, some of which was supported by the National Science Foundation.

We are here today because a Chinese research group recently published a paper in which they used these technologies to try to modify human embryos. That paper highlights scientific and ethical issues with these technologies, especially if they are being used to modify human germline cells as opposed to adult somatic cells.

I look forward to hearing about the science behind these technologies, as well as how the United States can be a leader in addressing the safety and ethical concerns associated with them.

I understand the National Academies has launched a major initiative around human gene editing technologies. In the 1970s, the National Academies played a similar role dealing with the then-new biotechnologies, and I look forward to hearing more about what they're planning to do concerning these new gene editing technologies. I also look forward to hearing about some of the potential nonhuman applications.

[The prepared statement of Mr. Lipinski follows:]

PREPARED STATEMENT OF SUBCOMMITTEE
MINORITY RANKING MEMBER DANIEL LIPINSKI

Thank you Chairwoman Comstock for holding this hearing on the science and ethics of new gene editing technologies. I want to thank all the witnesses for being here this afternoon and I look forward to hearing your testimony.

Although we are talking about gene editing technologies that are very new, it is important to mention that humans have been altering the genomes of species through selective breeding for thousands of years. Since the 1970s, it has been possible to directly manipulate DNA, which led to a biotechnology revolution and significant economic growth. Then we had the Human Genome Project to sequence the human genome that was coordinated by the Department of Energy and the National Institutes of Health. The full human genome was sequenced in 2003, opening up whole new possibilities for diagnosing and treating diseases. One such pathway led to the invention of the CRISPR technology.

Thanks to new gene editing technologies, which include CRISPR, we are able to add, remove, and replace DNA bases. They can be thought of as "search and replace" tools for DNA. They are incredibly powerful technologies that have the potential to transform the health care, energy, and agricultural sectors. Although new, these technologies were the outgrowth of decades of fundamental research, some of which was supported by the National Science Foundation. We are here today because a Chinese research group recently published a paper in which they used these technologies to try to modify human embryos. That paper highlights scientific and ethical issues with these technologies, especially if they are being used to modify human germline cells as opposed to adult somatic cells.

I look forward to hearing about the science behind these technologies as well as how the United States can be a leader in addressing the safety and ethical concerns associated with them. I understand that the National Academies has launched a major initiative around human gene editing technologies. In the 1970s, the National Academies played a similar role dealing with the then-new biotechnologies and I look forward to hearing more about what they are planning to do concerning these new gene editing technologies. I also look forward to hearing about some of the potential non-human applications.

Now I would like to yield my remaining time to my colleague from Illinois, Mr. Foster, who is very interested in this topic and helped organize today's hearing.

Mr. LIPINSKI. With that, I'd like to yield my remaining time to my colleague and neighbor from Illinois, Dr. Foster, who was very interested in this topic and helped to organize today's hearing.

Mr. FOSTER. Thank you, and I'd like to thank the whole Research and Technology Subcommittee, including the Chair, Congresswoman Comstock, and Ranking Member Lipinski for allowing me to join you here today. And similarly, I wanted to thank Chairman Smith and Ranking Member Johnson for agreeing to hold this hearing. And a very special thank you to the witnesses for taking their time out for this very important issue.

It is rare that prominent members of the scientific community come together to warn our leaders of technological breakthroughs that our legal system and society may not be prepared for, and yet, this is exactly what appears to be happening with recent discoveries in genetic editing tools.

As the last Ph.D. scientist in Congress, I am afraid I've served a sort of a lightning rod for many of these warnings and I take them very seriously.

I want to commend the National Academy of Sciences and the Institute of Medicine for the launch of their major initiative on human gene editing, and I want to make sure that Congress does everything constructive that it can to make sure that this is handled responsibly.

There is the possibility of very great benefits from these new technologies, and what makes them really revolutionary is what they can mean for humans, for example, replacing bone marrow of someone suffering from sickle cell disease with a modified version of their own marrow with the genetic defect removed.

However, if genetic modifications are made to so-called germline cells—these are sperm, eggs, embryos—then the modifications will be carried forward to future generations, which has implications that we need to carefully consider. We're on the verge of a technological breakthrough that could change the future of mankind and we must not blindly charge ahead.

Thank you, and I yield back my time.

Mr. LIPINSKI. I will just conclude. I agree with Dr. Foster and it's great to see that we have so many people here at this hearing. And it's a very important issue that we really need to consider deeply, so I thank the Chairwoman and the Chairman of the Full Committee, Chairman Smith, for holding this hearing today, and I'll yield back.

Chairwoman COMSTOCK. Thank you.

And I now recognize the Chairman of the Full Committee, Mr. Smith.

Chairman SMITH. Thank you, Madam Chair.

I do look forward, as do the others, to today's discussion on a new development in biology, which has been called "a game changer," "revolutionary," "powerful," and "a major issue for all humanity."

The new discoveries in genetically engineering human DNA offer potential cures for devastating genetic disorders. But the speed at which these new, simpler, and cheaper technologies are being used in the lab also presents ethical and health concerns. Most of the scientific community members have been clear: the science and ethics of this new technology must be resolved in order to prevent dangerous abuses and unintended consequences.

A recent report from China, where teams of researchers have begun to experiment with engineering DNA in human embryos, is alarming. This is an area where the United States can and should provide scientific and moral leadership. We need to better understand the technology and procedures being used so that we can ensure patients are treated in the safest and most ethical manner possible.

An April editorial in Science magazine called for a prudent path forward for genomic engineering. It recommended a moratorium on further research, while creating public forums for scientists, ethicists, and policymakers to discover the—to discuss "the attendant ethical, social, and legal implications of genome modification." This is why it is important that the House Science Committee is holding the first Congressional hearing on this profound and complex subject.

The purpose of the Science Committee is to explore the significance of scientific discoveries, as well as their potential implications for humankind. But we also must always be conscious of the potential ethical and moral issues raised by previously unimagined scientific breakthroughs. We must take the lead in reviewing new and innovative areas of science, such as genetically engineered DNA.

So I look forward, Madam Chair, to this informative discussion today, and we have an excellent panel of witnesses to hear from as well.

And I'll yield back.

[The prepared statement of Chairman Smith follows:]

PREPARED STATEMENT OF COMMITTEE ON SCIENCE, SPACE, AND TECHNOLOGY CHAIRMAN LAMAR SMITH

Thank you Madam Chair. I look forward to today's discussion on a new development in biology, which has been called "a game changer," "revolutionary," "powerful," and "a major issue for all humanity."

The new discoveries in genetically engineering human DNA offer potential cures for devastating genetic disorders. But the speed at which these new, simpler and cheaper technologies are being used in the lab also presents ethical and health concerns.

Most of the scientific community members have been clear: the science and ethics of this new technology must be resolved in order to prevent dangerous abuses and unintended consequences.

A recent report from China, where teams of researchers have begun to experiment with engineering DNA in human embryos, is alarming. This is an area where the United States can and should provide scientific and moral leadership.

We need to better understand the technology and procedures being used so that we can ensure patients are treated in the safest and most ethical manner possible.

An April editorial in Science Magazine called for a prudent path forward for genomic engineering. It recommended a moratorium on further research, while creating public forums for scientists, ethicists and policy makers to discuss "the attendant ethical, social, and legal implications of genome modification."

This is why it is important that the House Science Committee is holding the first congressional hearing on this profound and complex subject.

The purpose of the Science Committee is to explore the significance of scientific discoveries as well as their potential implications for humankind.

But we also must always be conscious of the potential ethical and moral issues raised by previously unimagined scientific breakthroughs.

We must take the lead in reviewing new and innovative areas of science, such as genetically engineered DNA.

I look forward to an informative discussion with our distinguished panel of witnesses.

Chairwoman COMSTOCK. Thank you.

And if there are Members who wish to submit additional opening statements, your statements will be added to the record at this point.

Now, at this time I would like to introduce our witnesses. Dr. Victor Dzau is the President of the National Academies Institute of Medicine and the James B. Duke Professor of Medicine at Duke University. Dr. Dzau has received many honors, including the Distinguished Scientist Award from the American Heart Association. He earned his undergraduate and medical degrees from McGill University and holds eight honorary doctorates.

Our second witness today is Dr. Jennifer Doudna. I think I got that. Dr. Doudna is Professor of Molecular and Cell Biology and Professor of Chemistry at U.C. Berkeley. A member of the National Academy of Sciences, Dr. Doudna is the recipient of several awards, including the NSF Waterman Award and the 2015 Breakthrough Prize for Life Sciences. Dr. Doudna earned her undergraduate degree in biochemistry from Pomona College and her Ph.D. in biological chemistry from Harvard University.

I now recognize the gentleman from Illinois, Mr. Lipinski, to introduce our next witness.

Mr. LIPINSKI. Thank you.

As a Northwestern University alumnus, I'm very excited to have Dr. McNally here today. To say an aside, Dr. Dzau, I'm also an alum of Duke University, and unfortunately for Dr. Doudna, an alum of Stanford also.

Dr. McNally is the Director of the Center for Genetic Medicine and Professor in the Departments of Medicine and Biochemistry at Northwestern University's Feinberg School of Medicine. She is a cardiologist who specializes in inherited forms of heart disease. Dr. McNally's research has identified genes and mechanisms for how genetic lead to heart and muscle disease. She has an undergraduate degree in biology and philosophy from Barnard College at Columbia University and an M.D. and Ph.D. from the Albert Einstein College of Medicine.

It is my pleasure to welcome Dr. McNally to our committee and look forward to her testimony.

Chairwoman COMSTOCK. Okay. And our final witness is Dr. Jeffrey Kahn, the Robert Henry Levi and Ryda Hecht Levi Professor of Bioethics and Public Policy at the Johns Hopkins Berman Institute of Bioethics and a Professor in the Department of Health Policy and Management at the Johns Hopkins Bloomberg School of Public Health. Dr. Khan received his bachelor's in microbiology from the University of California, Los Angeles, his master's in public health from Johns Hopkins, and his Ph.D. in philosophy and bioethics from Georgetown University.

In order to allow time for discussion, we would ask that you limit your testimony to five minutes and your entire written statement will be made part of the record.

I now recognize Dr. Dzau for five minutes to present his testimony.

TESTIMONY OF DR. VICTOR J. DZAU, PRESIDENT, INSTITUTE OF MEDICINE, THE NATIONAL ACADEMY OF SCIENCES

Dr. DZAU. Good afternoon, Chairman Smith, Chairwoman Comstock, Ranking Member Lipinski, and Subcommittee Members. As you heard, I'm Victor Dzau. I'm the President of the Institute of Medicine, which will soon be named the National Academy of Medicine on July 1.

I'm pleased to be here on behalf of the National Academies of Sciences, Engineering, and Medicine. The Academies operate under a Congressional Charter signed by Abraham Lincoln to provide advice to the Nation on matters where science, technology, and medicine can solve complex challenges and thereby improve people's lives.

Thank you for the opportunity to speak with you today about this very important matter of human gene editing and the major initiative that we have at the National Academies launched to help guide decision-making in this area.

The Academies have an established record of leadership on advising on emerging and often controversial areas of science, particularly genetic research, such as recombinant DNA and stem cell research. Our initiative is marshaling the best scientific evidence, medical, ethical, legal, and other expertise to help you and the Nation obtain a thorough understanding of gene editing and its potential risks and benefits. Our work is intended to provide a solid foundation to help inform decisions and policies on this research.

As you will hear from other witnesses today, gene editing technology holds great promise. In fact, powerful new tools such as CRISPR/Cas9 developed by my colleague Dr. Doudna and others, as well as other genetic engineering technology, allow researchers now to precisely modify the genetic makeup of any living organism, including humans. The possible benefit application for such technologies are many. They could offer a cure to devastating genetic diseases such as Huntington's disease, as you heard, or sickle cell anemia. It can help improve and understand the treatment of many other illnesses.

These technologies also present complex challenges both to scientific and medical communities and to society as a whole. Of particular concern is the potential to make permanent modification to human DNA in nuclei of eggs, sperm, or human embryos that are then passed down to succeeding generations known as germline gene editing. Research that attempts to alter the human germline raises important issues in so many different ways about safety, risk, social, economic, ethical, and regulatory considerations. So although more remains to be done before these technologies can be deployed safely, their availability certainly intensifies this debate among scientists and physicians about such research.

Here in the United States there are legislative prohibitions on the use of federal funds for research of human embryos and there exist constraints on such research when subject to oversight by the U.S. Food and Drug Administration or other government agencies. These constraints, however, do not apply to work done internationally without federal funds and without the intent to seek federal approval of any product of that research. So clearly, we have reached a critical juncture in genetic editing research and guidance is needed.

The National Academies of Sciences, Engineering, and Medicine are prepared to provide that guidance based on an in-depth review of science underlying gene editing, the potential benefits, and the valid concerns raised by this research. Toward that end, our initiative on human gene editing research is already underway. Just last week, we held the first meeting of a multidisciplinary advisory group that will help us steer our initiative and ensure the Academies' efforts are comprehensive, inclusive, and transparent.

Since much of this research is done internationally, the Academies will convene a global summit to obtain multinational perspectives on recent scientific development in human gene editing and the associated ethical and governance issues. Concurrently, we'll appoint an expert committee to conduct a comprehensive study of the scientific underpinning and clinical, ethical, legal, and social implications of human gene editing. We hope that we will come up with recommendations which can inform the Nation for decisions in this area.

All of us—scientists, physicians, policymakers, and public—want to do everything possible to ensure the scientific and medical breakthroughs benefit all of mankind and do no harm. The Academies are certainly ready to help.

I would be very happy to answer any questions the Subcommittee may have. Thank you.
[The prepared statement of Dr. Dzau follows:]

"An Initiative to Guide Decision Making on Human Gene-Editing Research"

Statement of Victor J. Dzau
President, National Academy of Medicine
National Academies of Sciences, Engineering, and Medicine

Before the

Subcommittee on Research and Technology
Committee on Science, Space, and Technology
U.S. House of Representatives

June 16, 2015

Good afternoon, Chairwoman Comstock, Ranking Member Lipinski, and members of the Subcommittee. I'm Victor Dzau, president of the Institute of Medicine, which will become the National Academy of Medicine on July 1. I'm pleased to be here today on behalf of the National Academies of Sciences, Engineering, and Medicine. The Academies operate under a congressional charter signed by Abraham Lincoln in 1863 to provide advice to the nation on matters where science, technology, and medicine can solve complex challenges and thereby improve peoples' lives.

Thank you for the opportunity to speak with you today about the important matter of human gene editing and the major initiative we have launched to help guide decision making in this area. The Academies have an established track record of providing leadership in emerging and often controversial areas of genetic research. Our initiative is marshalling the best available expertise to help you and the nation obtain a thorough understanding of gene editing and its potential benefits and risks, which will provide a solid foundation for informed decisions and sound policies on this research.

Potential Benefits and Challenges

As you will hear from other witnesses today, gene-editing technologies hold great promise for advancing science and improving human health. Powerful new tools such as CRISPR-Cas9 developed by Dr. Doudna and others allow researchers with basic knowledge of molecular genetics to precisely modify the genetic makeup of any living organism. The possible applications for such technologies are many. The genomes of plants and animals could be modified to boost agriculture and food production. Genes of disease-carrying insects could be edited to reduce the spread of malaria, West Nile virus, or dengue fever. In humans, the technologies could offer a cure to often devastating genetic diseases such as Huntington's disease and sickle cell anemia, and help improve understanding and treatment of many other illnesses.

However, these new avenues of research also present many complex challenges, both to the scientific and medical communities and to society as a whole. Research that attempts to alter human genes raises important questions about safety, uncertainties, risks, and ethical considerations. Of particular concern is the potential to make permanent modifications to human DNA in the nuclei of cells in eggs, sperm, or human embryos that are then passed down to succeeding generations. This is known as human germline editing.

Although much remains to be done before these technologies could be deployed safely and efficiently, the availability of these new technologies has certainly intensified debate among scientists and physicians about such research. Recently, through an editorial in a prominent scientific journal, several researchers, including Dr. Doudna, have called for a suspension of studies that attempt to modify the human germline for clinical application until the potential risks, benefits, and ethical concerns are thoroughly explored.

Here in the U.S., there are legislative prohibitions on the use of federal funds for research on human embryos, and constraints on such research when it is subject to oversight by government agencies such as the U.S. Food and Drug Administration. These laws and regulations do not apply to work done internationally without federal funds and without the intent to seek federal approval of any products of that research. In April, Francis Collins, the director of the National Institutes of Health, stated that NIH "will not fund any use of gene-editing technologies in human embryos." Collins cited strong arguments against such research, including "serious and unquantifiable safety issues, ethical issues presented by altering the germline in a way that affects the next generation without their consent, and a current lack of compelling medical applications justifying the use of CRISPR/Cas9 in embryos." And in May, John Holdren, the director of the White House Office of Science and Technology Policy, said that "the Administration believes that altering the human germline for clinical purposes is a line that should not be crossed at this time."

The Academies Initiative

It's clear that the advent of these technologies has brought us to a critical juncture in genetic research. What is needed now is guidance – guidance that is based on an in-depth review of the science underlying gene editing and an understanding of the potential benefits as well as the valid concerns raised by this research. This is exactly the type of leadership for which the National Academies of Sciences, Engineering, and Medicine are known.

Toward that end, on May 18, we announced a major initiative on human gene-editing research. Our work is already well-underway. Just last week, we met with a multidisciplinary advisory group that will help steer our initiative. Their names are appended to my testimony. This group will be instrumental in counseling Ralph Cicerone, president of the National Academy of Sciences, and me to ensure that the Academies' efforts in this area are comprehensive, inclusive, and transparent.

As with science and medicine in general, gene-editing research is truly an international endeavor, and any future applications will likely be felt around the world. To gather the multinational, multidisciplinary perspectives critical to the success of this initiative, the Academies will convene a global summit to examine recent scientific developments in human gene editing and the range of associated ethical and governance issues. Summit participants will examine:

- The current state of the science and available technologies;
- The rationale for conducting gene-editing research in humans;
- Existing national and international regulatory principles, standards, and guidance for such research and areas where more direction is needed; and
- Ethical and legal considerations in such research.

Concurrently, the Academies will appoint an expert committee to conduct a comprehensive study on human gene-editing research. Like all of our committees, this study committee will represent a wide range of expertise and be carefully screened for bias and conflict of interest. Although the study's statement of task is still being finalized, some of the questions the committee will likely address include:

- What is the state of the science of gene editing and how rapidly is it advancing?
- What is the evidence on the efficacy and risks of gene editing in humans?
- What are the potential clinical applications and how should their risks and benefits be weighed for current and future generations?
- What principles and frameworks should be applied for determining which, if any, applications should go forward?
- What are the ethical, legal, and social implications?

- What oversight mechanisms are needed and which safeguards should be in place to guard against misuse of gene-editing techniques?

Of course, the study will also be informed by our international summit.

Advances to Benefit Humankind

I am confident that the Academies' initiative will help the nation and the world make sound, evidence-based decisions about this research. Allow me to briefly highlight a few examples of when the Academies have been of similar service

In 1975, the National Academy of Sciences convened what is now known as the Asilomar conference, a landmark turning point for recombinant DNA research. The conference ultimately led to voluntary guidelines to ensure the safety of what was then a new technology. Our 1988 study on mapping the human genome helped steer what has become an incredible source of new scientific advances. In 2005, we issued guidelines for human embryonic stem cell research, which were widely adopted by research institutions, and international scientific societies. And most recently, an international workshop on research of dangerous pathogenic viruses – known as "gain of function" research – will inform new policies for the study of avian influenza, Severe Acute Respiratory Syndrome (SARS), and Middle East Respiratory Syndrome (MERS).

These examples underscore how quickly genetics and biomedical science have advanced over just the past few decades. It's no wonder that all of us – scientists, physicians, policymakers, and the public – want to do everything possible to ensure that these advances continue and that scientific and medical breakthroughs such as gene editing benefit all of humankind. With that goal in mind, the National Academies of Sciences, Engineering, and Medicine are ready to provide a comprehensive understanding of human gene editing and its implications to help guide decisions about its use in the years to come.

Thank you for inviting me to testify. I would be pleased to address questions from the Subcommittee.

Biography of Victor Dzau, M.D.

Victor J. Dzau is the eighth President of the Institute of Medicine (IOM). Dr. Dzau is Chancellor Emeritus and James B. Duke Professor of Medicine at Duke University and the past President and CEO of the Duke University Health System. Previously, Dr. Dzau was the Hersey Professor of Theory and Practice of Medicine and Chairman of Medicine at Harvard Medical School's Brigham and Women's Hospital, as well as Chairman of the Department of Medicine at Stanford University.

Dr. Dzau has made a significant impact on medicine through his seminal research in cardiovascular medicine and genetics and his leadership in health care innovation. His important work on the renin angiotensin system (RAS) paved the way for the contemporary understanding of RAS in cardiovascular disease and the development of RAS inhibitors as widely used, lifesaving drugs. Dr. Dzau also pioneered gene therapy for vascular disease, and his recent work on stem cell paracrine mechanisms and the use of microRNA in direct reprogramming provides novel insight into stem cell biology and regenerative medicine. In his role as a leader in health care, Dr. Dzau has led efforts in innovation to improve health, including the development of the Duke Translational Medicine Institute, the Duke Global Health Institute, the Duke-National University of Singapore Graduate Medical School, and the Duke Institute for Health Innovation.

As one of the world's preeminent health leaders, Dr. Dzau advises governments, corporations, and universities worldwide. He has served as a member of the Advisory Committee to the Director of the National Institutes of Health (NIH) and as Chair of the NIH Cardiovascular Disease Advisory Committee. Currently he is a member of the Board of the Singapore Health System and Hamad Medical Corporation, Qatar. He was on the Board of Health Governors of the World Economic Forum and chaired its Global Agenda Council on Personalized and Precision Medicine.

Among his many honors and recognitions are the Gustav Nylin Medal from the Swedish Royal College of Medicine, the Distinguished Scientist Award from the American Heart Association, Ellis Island Medal of Honor, and the Henry Freisen International Prize. In 2014, he received the Public Service Medal from the President of Singapore. He is a member of the Institute of Medicine of the National Academy of Sciences, the American Academy of Arts and Sciences and the European Academy of Sciences and Arts. He has received eight honorary doctorates.

Chairwoman COMSTOCK. And Dr. Doudna.

**TESTIMONY OF DR. JENNIFER DOUDNA,
PROFESSOR OF BIOCHEMISTRY AND MOLECULAR BIOLOGY,
UNIVERSITY OF CALIFORNIA, BERKELEY**

Dr. DOUDNA. Good afternoon, Chairwoman Comstock and the rest of the Members of the Committee. It's a great pleasure for me to be here and have the opportunity to talk with you about science that I've been involved with from its origin and involved in leading the discussion of where it's going.

I wanted to start by saying that this is research that originated as a basic science project funded in part by the National Science Foundation. We did not aim to develop a genome editing technology but in the course of the experiments that we were doing, it became clear that what started as a study of a bacterial immune system, the way bacteria fight the flu, could actually be reengineered and re-harnessed really as a technology for changing sequences in the genomes of cells and whole organisms.

I wanted to tell you a little bit more about the science behind this to explain a little bit about how it works and why it's revolutionary. So I think what really makes this distinct from other ways of manipulating DNA and cells is that it's a very simple system. It relies I would really make the analogy to software that you use for your computer. Here we have a protein called Cas9 that can be easily reprogrammed by using a short piece of nucleic acid called RNA that enables this protein to be directed to essentially any DNA sequence in a genome of an organism. And because genome sequencing has become very prevalent and is becoming less and less expensive, we have an exciting convergence of technologies that give us information about the entire genome in a cell or an organism and now a tool that allows scientists to change that sequence in a very precise fashion so we can do things, as was mentioned in the opening statements, like correct mutations that would otherwise lead to genetic disease.

So this is, I think, a very exciting moment in biology. It's opened up a lot of opportunities for research, for clinical applications in the future, but it also raises various questions about the way that this technology should be employed going forward, and in particular in our discussion today the question of whether and when this technology should be employed to change the sequence in the human germline in eggs or sperm or embryos that would lead to a genetically modified person that would be a mutation that could be passed down to their children.

And I realized fairly early on in our research that this technology was likely to be applicable in the human germline, and that led me to initiate a discussion initially with some—a fairly small group of scientists in California. We met in January of this year in the Napa Valley to discuss this very issue and we spent a day. We had—that small meeting included scientists, clinicians, as well as bioethicists to discuss the various issues around human germline editing, and that meeting resulted in a perspective that was published in Science magazine about two months ago that was referred to in the opening statements for—that called for a prudent path forward in any kind of clinical application of germline editing in humans.

I do want to point out that our perspective favors research in this area. I think we feel as scientists that it's very important to have data so that we can make informed decisions about future potential applications, and I think this is a—I think many of us appreciate this is a technology that could be very helpful for people that have inherited genetic disorders. However, to ensure that any kind of application clinically in the human germline was safe and really was used in an ethical fashion, we do need to understand how this technology operates in those types of cells.

Thank you.

[The prepared statement of Dr. Doudna follows:]

The Science and Ethics of Human Genome Engineering

Overview and description of CRISPR-Cas9 technology

The rapidly expanding family of CRISPR-Cas9-derived technologies is revolutionizing the fields of genetics and molecular biology as researchers worldwide employ these methods to change DNA sequences – by introducing or correcting genetic mutations – in a wide variety of cells and organisms. The simplicity and efficiency of the CRISPR-Cas9 system enables any researcher with knowledge of molecular biology to modify genomes, making feasible many experiments that were previously difficult or impossible to conduct. The technology employs a bacterial protein, Cas9, that cuts specific DNA sequences defined by base pairing between a guide RNA molecule and a DNA target sequence. Researchers can easily alter the guide RNA sequence to direct Cas9 to desired sites in the genome of cells. The resulting double-stranded DNA break triggers site-specific sequence changes, enabling disruption or recoding of genes at will. Where older technologies were "hard-wired", requiring new proteins to be engineered for each experiment, the CRISPR-Cas9 system is analogous to software that is easily reprogrammable for a wide variety of experiments and functions across a broad range of plant and animal systems. For example, the CRISPR-Cas9 system enables introduction of DNA sequence changes that correct genetic defects in whole animals and cultured tissues produced from stem cells, strategies that could eventually be used to treat human disease. This technology can also be used to replicate precisely the genetic basis for human diseases in model organisms, leading to unprecedented insights into previously enigmatic disorders. For more information, see videos and references posted at www.innovativegenomics.org.

In addition to facilitating changes in differentiated somatic (adult) cells of animals and plants, CRISPR-Cas9 technology can also be used to change the DNA in the nuclei of reproductive cells that transmit information from one generation to the next (an organism's "germ line"). Thus, it is now possible to employ CRISPR-Cas9 for genome modification in fertilized animal eggs or embryos, thereby altering the genetic makeup of every differentiated cell in an organism and thus ensuring that the changes will be passed on to the organism's progeny. Humans are no exception - changes to the human germ line are now possible using this simple and widely available technology.

A prudent path forward for human germ line modification

The possibility of human germ line engineering has long been a source of both excitement and unease among the general public, especially in light of concerns about initiating a "slippery slope" from disease-curing applications toward uses with less compelling or even troubling implications. A key point of discussion is whether the treatment or cure of severe diseases in humans would be a responsible use of germ line genome engineering, and if so, under what circumstances. For example, would it be appropriate to use the technology to change a disease-causing genetic mutation to a sequence more typical among healthy people? Even this seemingly straightforward scenario raises serious concerns, including the potential for unintended consequences of heritable germ line modifications, since there are limits to our knowledge of human genetics, gene-environment interactions and the pathways of disease (including the interplay between one disease and the other conditions or diseases in the same patient).

In the United States, human research currently requires an Investigational New Drug (IND) exemption from the Food and Drug Administration, but some countries do not have such regulations in place. A recent Perspective that I co-authored advocated a halt to any clinical applications of human germ line editing until the safety, efficacy and ethical considerations of such use have been assessed (Baltimore et al. (2015) *Science*).

Legal and societal concerns around genome engineering

The CRISPR-Cas9 technology makes it easier and faster to introduce site-specific changes into the DNA of cells, tissues and whole organisms on a scale far beyond what has been possible in the past. Anyone with basic knowledge of molecular biology can employ this powerful technology, which makes regulation difficult. In addition, the versatility of the technology raises various scenarios that warrant careful consideration, review and oversight. Applications of genome engineering that create heritable changes in humans, introduce self-propagating mutations that sterilize insects or trigger chromosome translocations that cause cancer are examples that have already been published in the scientific literature and could raise both legal and societal concerns.

Towards a responsible framework for using genome editing technologies

The importance and complexity of the issues surrounding some applications of genome engineering methods warrant thoughtful review with a goal of crafting guidelines for responsible use. Towards this end, the National Academy of Sciences and the National Academy of Medicine are co-sponsoring forthcoming meetings and will commission formal reports to provide guidance to scientists, clinicians and regulatory agencies. These meetings will involve international participation by people representing different scientific organizations and points of view. Past experience with technologies such as molecular cloning, *in vitro* fertilization and embryonic stem cell manipulation should help inform the course of action taken.

US role in scientific and ethical leadership in this area of science

The US has three critical roles to play in creating a responsible framework for the use of genome engineering technologies. The first is to provide expert information and recommendations to the scientific community about the risks and benefits of genome engineering for various types of applications in humans, other mammals, and organisms including plants, insects and microbes. The second is to lead an international consortium of scientists and clinicians in drafting guidelines for use that will form the basis for regulation and oversight by governments worldwide, particularly for applications involving the human germ line. The consortium should also provide advice about other applications of genome editing that could impact the environment, food security and human health. And the third role the US should play is to educate the public about the benefits and risks of genome editing. This will enable non-scientists to understand the opportunities as well as the potential dangers of the technology and to make decisions about its use from an informed perspective.

Brief biography of Jennifer Doudna

Jennifer Doudna is the Li Ka Ching Chancellor's Chair in Biomedical and Health Sciences and she is Professor of Molecular and Cell Biology and Professor of Chemistry at UC Berkeley and an Investigator of the Howard Hughes Medical Institute. She received her undergraduate degree in Biochemistry from Pomona College in 1985 and her Ph.D. in Biological Chemistry from Harvard University in 1989. After completing postdoctoral work at the University of Colorado at Boulder, she joined the Yale University faculty in 1994, where she was promoted through the ranks to Henry Ford II Professor of Molecular Biophysics and Biochemistry in 1999. In 2002 she joined the faculty at UC Berkeley. Prof. Doudna is a member of the National Academy of Sciences, the American Academy of Arts and Sciences, the Institute of Medicine and the National Academy of Inventors. She is a recipient of awards including the NSF Waterman Award, the FNIH Lurie Prize, the Paul Janssen Award for Biomedical Research, the Breakthrough Prize in Life Sciences, the Princess of Asturias Award (Spain) and the Gruber Prize in Genetics.

Chairwoman COMSTOCK. Thank you.
Now, we'll hear from Dr. McNally.

**TESTIMONY OF DR. ELIZABETH MCNALLY,
PROFESSOR OF GENETIC MEDICINE,
PROFESSOR IN MEDICINE-CARDIOLOGY AND BIOCHEMISTRY
AND MOLECULAR GENETICS;
DIRECTOR, CENTER FOR GENETIC MEDICINE,
NORTHWESTERN UNIVERSITY**

Dr. MCNALLY. Thank you.

On behalf of Northwestern University, I'd like to thank Chairwoman Comstock and Ranking Member Lipinski for inviting me here today.

I'm Elizabeth McNally. I'm the Ward Professor of Genetic Medicine, and I direct the Center for Genetic Medicine at Northwestern. I'm a cardiologist and I specialize in providing care for patients and families with inherited forms of heart disease. Over the last decade, we've seen a dramatic increase in available genetic testing and we now routinely provide genetic diagnosis, risk assessment, and importantly, risk reduction for genetic diseases that affect the heart.

Diseases like cystic fibrosis, Duchenne muscular dystrophy, sickle cell anemia are those that are caused by mutations in single genes. The gene editing techniques that we are here discussing offer the opportunity for permanent lifelong treatment of those disorders. With advances in DNA sequencing technology, it is now possible to sequence an individual genome in a day. For less than the cost of an MRI, a genome can be analyzed with high accuracy pinpointing single gene mutations. The Office of Rare Diseases identifies nearly 7,000 rare diseases and many of these are genetic in origin, often arising from single mutations. The ORD estimates that nearly 30 million Americans are affected by rare diseases. More than half of rare diseases affect children.

Concomitant with advances in genetic diagnoses, there are parallel leaps in genome editing. CRISPR/Cas9 represents a significant advance for genome editing. Because of the co-development of gene editing and stem cell biology, there is significant potential clinical application. Induced pluripotent human stem cells can be made from blood, skin, and other mature human cells. For my field, cardiology, skin cells can actually form beating heart-like cells in a dish allowing us to discern how mutations cause disease and letting us test how to correct these diseases. The human population is not placed at risk by these experiments in cells and it seems fair to say that the human population would actually be harmed by not doing these experiments since the research offers the potent opportunity to improve human health.

Stem cells of the bone marrow, muscle, skin, and other organs can be isolated and edited. With these methods, it would be possible to cure sickle cell anemia or Duchenne muscular dystrophy. In mice, CRISPR/Cas9 mediated correction of fertilized eggs corrected the defect for Duchenne muscular dystrophy. The method, while imperfect, was associated with a remarkably high correction rate.

Recently, a group of distinguished scientists called for careful consideration of gene editing in fertilized oocytes fearing the potential for germline gene editing and ultimately human eugenics. These discussions were enhanced and prompted by the recent report which we've heard about from Liang, et al., which described the efforts using CRISPR/Cas9 in fertilized oocytes.

A regulatory framework for gene editing should encompass several key points. It should permit research to optimize and improve CRISPR/Cas9 and related technologies. It should permit in vitro and cell-based gene editing technologies, including those in embryonic and induced pluripotent stem cells. It should permit in vitro and cell-based editing with the intent to reintroduce cells into humans as a therapeutic measure for somatic cells. And it should permit the generation of gene-edited animals for the purposes of scientific research. It may consider limiting or even prohibiting gene editing under the circumstances where human transmission of gene-edited germ lines would occur.

But would we ever really consider germline gene editing? So we should consider the scenario of pre-implantation genetic diagnosis, otherwise known as PGD. PGD is pursued by families to avoid transmitting genetic diseases. PGD involves in vitro fertilization coupled with genetic testing. PGD is not covered by insurance, and yet for some families, they make this choice. These may be families who are already struggling with caring for one disabled child and who cannot care for a second disabled child.

PGD is a personal option and one that is made solely by parents and families. In principle, it is possible that the efficiency of genome editing will improve so that pre-implantation genetic correction could accompany PGD. With this process, gene editing to correct and eliminate a genetic disease could become reality. While the temptation may be to ban or limit this possibility, we should do so only with caution. Parents of children with genetic disease express significant concern, responsibility, and often dismay for having passed on mutations to their children. A parent's desire to protect children is undeniable. As a society and as a nation, we protect children.

It may be tempting and easiest to ban gene editing where germline transmission could occur, yet the justified use of this approach is certainly conceivable and may one day be appropriate.

Thank you.

[The prepared statement of Dr. McNally follows:]

TESTIMONY

OF

PROFESSOR ELIZABETH MCNALLY
Northwestern University Feinberg School of Medicine
Chicago, Illinois

to the

Committee on Science, Space, and Technology
Subcommittee on Research and Technology
Hearing on
The Science and Ethics of Genetically Engineered Human DNA
U.S. House of Representatives
16 June 2015

INTRODUCTION

On behalf of Northwestern University, I would like to thank Chairwoman Comstock and Ranking Member Lipinski for inviting me here today to speak at this hearing on Genetically Engineered Human DNA. I would like to also thank the Subcommittee for convening this hearing.

I am the Elizabeth J. Ward Professor of Genetic Medicine, and I direct the Center for Genetic Medicine at Northwestern. I am a cardiologist who specializes in providing care for patients and families with inherited forms of heart disease. I established one of the first Cardiovascular Genetics Clinics in the United States. Over the last decade, we have seen a dramatic increase in available genetic testing, and we now routinely provide genetic diagnosis, risk assessment, and risk reduction of genetic diseases that affect the heart. Many of the inherited diseases that we diagnose and manage are also those that affect muscle. The same gene mutations that cause heart muscle to weaken may elicit the same effect on skeletal muscle, causing those who carry the mutations to develop heart failure, life threatening irregular heart rhythms and muscle weakness. Genetic diagnosis is not restricted to heart and muscle disorders as nearly every area of medicine is influenced by genetic diagnosis.

Genetic Diseases

Diseases like Cystic Fibrosis, Duchenne Muscular Dystrophy, and Sickle Cell are those that are caused by mutations in single genes. It has been possible for some time to genetically diagnose these disorders. There is considerable effort directed at devising targeted therapies to correct the underlying genetic defects responsible for causing disorders like these, and herein I will discuss the potential application of gene editing techniques for the treatment of genetic diseases.

The first draft human genome sequence was completed just 15 years ago. Now, with advances in DNA sequencing technology it is now possible to sequence an individual human genome in a matter of days. Moreover, human genome sequencing can be completed a comparatively low cost, less than the cost of an MRI, and can be analyzed with high accuracy. It is becoming routine to pinpoint single gene mutations responsible for devastating disorders, including those diseases that affect children. With this explosion in genetic analysis, the number of genetic disorders is increasing. The National Institutes of Health (NIH) Office of Rare Diseases identifies nearly 7000 rare diseases[1], and many of these are genetic in origin, often arising from single mutations. The ORD estimates that nearly 30 million Americans are affected by rare diseases. More than half of rare diseases affect children.[2]

Gene Editing

Concomitant with advances in genetic diagnoses, there are parallel leaps in genome editing. The concept of genome editing is not new. Genome editing has been technically possible since the first reports of inserting genetic material in fertilized eggs of mice, reported by three independent groups in 1981.[3-5] It was around this very same time that the first successes in human in vitro fertilization were reported.[6] In the more than three decades since genetic editing became possible, there has been scientific and technological progress that has improved the proficiency and fidelity of genome editing. Early success in genomic editing relied on random insertion of new DNA sequences into fertilized eggs, stem cells, and cell lines. Random insertion allows new genes to be expressed but does not correct genetic defects. Homologous recombination refers to the process by which sequences can be exchanged between a vector that carries new sequences of insert and the genome to be edited. For most organisms, especially humans, homologous recombination is a remarkably inefficient process. However in other organisms, homologous recombination occurs at much higher rate. Understanding the precise means by which organisms can alter genetic structures has allowed researchers to isolate the machinery that edits genomes. In the last decade, there have been several key discoveries made to improve the ability to precisely change specific sequences. The precision of gene editing remains at the center of these discussions. Precise gene editing refers to producing the desired genetic change, and importantly doing so with high efficiency and with few off target effects.

The most recent advance capitalizes on the tools used by bacteria to ward off viral infection. This newest technology, referred to as CRISPR/Cas9, isolates the sequences and enzymes used by bacteria, and then applies these methods into complex cell types like those in mice, rats and humans.[7] First described in 2012, CRISPR/Cas9 is changing the path and pace of scientific discovery. Research depends on model systems, and model systems include cultured cells, as well as organisms like mice, yeast, flies, worms and other species. Genetically tractable systems are preferred, and mice remain a standard for the field of human biology. Cell models of disease are also highly useful since experiments can be completed with comparative ease and speed. The timeline of discovery is tied tightly to the model system of choice. The importance of CRISPR/Cas9 cannot be overstated. CRISPR/Cas9 offers a precision heretofore unseen. Cells and animals can be manipulated to more precisely to facilitate the ability to ask and answer critically important scientific questions.

Gene editing in Stem Cells

Alongside these advances in genome editing, it is worthwhile to consider gains in stem cell biology. The application of genome editing goes hand in hand with stem cell biology, and because of this co-evolution of gene editing and stem cell biology, there is significant potential clinical application. The ethics, merits, and implications of human embryonic stem cell biology have been debated and will not be reiterated here. For the purposes of this testimony, it should be acknowledged that some human stem cell lines retain the ability to contribute to human germ cells. In contributing to human germ cells, there is the possibility to transmit stem cell-derived genetic material into new generations. Therefore, genome editing in certain stem cells, in principle, may have the ability to alter human germ lines. However, many stem cells only theoretically have the capacity to contribute to human germ lines. In practice, human stem cells are used in many experiment with no intent or possibility of contributing to human germ lines. Induced pluripotent human stem cells can be made from blood, skin and other mature somatic human cells. Induced pluripotent stem cells, in theory, could contribute to human germ lines but are not used for this purpose. In many laboratories, the true stem cell capacity of such stem lines is never evaluated, even in mice, as the germ line potential is irrelevant to the research.

Embryonic and induced pluripotent stem cells are an obvious venue in which to test and evaluate genome editing techniques. The value of stem cells lines is that we can study how mutations act in many different cell types. Cells can be induced to form beating heart-like cells in a dish. How a disease-causing mutation affects beating and function can now be readily understood in cell culture. Introducing new mutations into stem cells generates highly valuable models for human disease. These models are then used to identify and test new therapies. The human population is not placed at risk by these experiments in cells. It seems fair to state that the human population would actually be harmed by not doing these experiments since this research offers a potent opportunity to improve human health. This is not an opportunity that should be missed. Having genome-edited cell lines allows more rapid scientific advance and reduces the need for certain types of animal experimentation. At the same time, correcting defective genes in stem cells allows investigators to determine whether genomic correction is possible. In principle, a corrected stem cell could prove useful in cell transplant experiments to treat some diseases.

Gene editing is not restricted to pluripotent stem cells. Stem cells of the bone marrow, muscle, skin and other organs and tissues can be isolated and edited. In this case, editing and correction could be accompanied by transplant into a host human in order to treat disease. With this method, it would be possible to cure Sickle Cell Anemia or Duchenne Muscular Dystrophy. At present, the methods CRISPR/Cas9 require optimization in order for this to be reality. But the advances of CRISPR/Cas9 bring this approach into discussion. In mice, CRISPR/Cas9-mediated correction in fertilized oocytes corrected the defect of Duchenne Muscular Dystrophy.[8] The method, while imperfect, was associated with remarkably high correction.

Germ line gene editing

Recently a group of distinguished scientists called for a moratorium on gene editing in human fertilized oocytes fearing the potential of germ line gene editing and, ultimately, human eugenics.[9] These discussions were enhanced and prompted by the recent report of Liang et al. described efforts using CRISPR/Cas9 gene editing in fertilized human zygotes.[10] To limit concerns regarding human eugenics, the authors used tripronuclear zygotes that are genetically limited from progressing through development into humans. Notably, the authors concluded that CRISPR/Cas9, while an improvement over previous gene editing technologies, still has limited efficiency and importantly has serious off-target effects. The major off-target effect is the introduction of unintended mutations at sites throughout the genome at an unacceptably high rate for clinical purposes. Whether CRISPR/Cas9's efficiency and off-target effects differ across cell types is not well known at present. However, these same issues are present in all cell types subjected to gene editing to date. With knowledge of the enzymes, sequences and structures of the CRISPR/Cas9 system, optimization is an active area of research in academic, government and private industry laboratories in the United States and throughout the world.

Regulating Gene Editing

A regulatory framework for gene editing should encompass several key points:

1) Permit research to optimize and improve CRISPR/Cas9 and related technology.

2) Permit in vitro, cell-based gene editing technologies, including those in embryonic and induced pluripotent stem cells, respecting regulations currently protecting human embryonic stem cell lines.

3) Permit in vitro, cell-based gene editing with the intent to re-introduce into humans as a therapeutic measure for somatic cells. For example, this would apply to gene edited bone marrow derived stem calls. The treatment of a human with a gene edited cells would fall under the existing regulatory framework.

4) Permit the generation of gene-edited animals for the purposes of scientific research.

5) Limit or prohibit gene editing under circumstances where human transmission of gene-edited germ lines would occur.

Why consider germ line gene editing?

With current technology, it is difficult to envision any justifiable use of gene editing in fertilized human zygotes where the resultant edited genome would be transmitted to future generations. Yet, we should consider the scenario of pre-implantation genetic diagnosis (PGD). PGD is pursued by families to avoid transmitting genetic diseases. Most commonly PGD is only pursued related to genetic diseases associated with significant early onset morbidity and mortality. With more widespread use of genetic diagnosis, as a clinician, I am asked about options to avoid passing deleterious genetic mutations to the next generation.

PGD involves in vitro fertilization coupled with genetic testing. In PGD, in vitro fertilization is used to create a fertilized oocyte that undergoes several rounds of cell division to become an embryo.[11] A single cell is removed from the embryo and tested genetically to identify those embryos that do not carry a specific genetic mutation. PGD allow parents to implant only those embryos free of the mutation in question. PGD is limited by the number of available embryos. PGD is typically not covered by insurance, and yet some families make this choice. These may be families who are already struggling with caring for one disabled child who cannot care for a second disabled child. These may be families where the parent is significantly afflicted with a genetic disease, and the parent wishes not to have his or her child burdened with the same diagnosis. PGD is a personal option and one that is made by solely by parents and families. PGD is not new and has been an available option for the last decade. A relatively small number of families choose this option and the choice to do so is often limited by technology, cost, religious and personal preference. PGD relies on nature to provide embryos free of a specific genetic mutation. Genetic altering of human embryos has occurred in the form of adding mitochondria from an external source, which introduces new mitochondrial DNA. In principle, it is possible that the efficiency of genome editing will improve so that preimplantation genetic correction could accompany PGD. With this process, gene editing to correct and eliminate a genetic disease could become reality. While the temptation may be to ban or limit this possibility, we should do so only with caution.

In my many years of working with patients and families with genetic disease, I can report that many parents of children with genetic disease express significant concern and responsibility for having passed on mutations to their children. A parent's desire to protect children is undeniable. As a society and as a nation, we embrace and endorse the importance of protecting children. It may be tempting, and perhaps easiest, to ban all gene editing where germ line transmission could occur. Yet, the justified use of this approach is certainly conceivable and may one day be appropriate.

REFERENCES

1. https://rarediseases.info.nih.gov/about-ordr/pages/31/frequently-asked-questions. Date accessed June 9, 2014.

2. http://globalgenes.org/rare-diseases-facts-statistics/ Date accessed June 9, 2015.

3. Gordon J, Ruddle F (1981) Integration and stable germ line transmission of genes injected into mouse pronuclei. *Science* **214** (4526): PMID 6272397.

4. Costantini F, Lacy E (1981) Introduction of a rabbit β-globin gene into the mouse germ line. *Nature* **294** (5836): 92–4. PMID 6945481.

5. Brinster R, Chen HY, Trumbauer M, Senear AW, Warren R, Palmiter RD (1981) Somatic expression of herpes thymidine kinase in mice following injection of a fusion gene into eggs. *Cell* **27** (1 pt 2): 223-231. PMID 6276022.

6. http://www.nytimes.com/2014/03/04/health/a-powerful-new-way-to-edit-dna.html?_r=0. Date accessed June 9, 2015.

7. Wang J, Sauer MV (2006) In vitro fertilization: a review of three decades of clinical innovation and technical advancement. *Ther Clin Risk Management* **2** (4): 355–364. PMID: 18360648, PMCID: PMC1936357.

8. Long C, McAnally JR, Shelton JM, Mireault AA, Bassel-Duby R, Olson EN (2014) Prevention of muscular dystrophy in mice by CRISPR/Cas9 mediated editing of germline DNA. *Science* **345** (6201) 1184-1181. PMID: 25123483, PMCID: PMC4398027.

9. Baltimore D, Berg P, Botchan M, Carroll D, Charo RA, Church G, Corn JE, Daley GQ, Doudna JA, Fenner M, Greely HT, Jinek M, Martin GS, Penhoet E, Puck J, Sternberg SH, Weissman JS, Yamamoto KR (2015) Biotechnology. A prudent path forward for genomic engineering and germline gene modification. *Science* **348** (6230) :36-8. PMID: 25791083, PMCID: PMC4394183.

10. Liang P1, Xu Y, Zhang X, Ding C, Huang R, Zhang Z, Lv J, Xie X, Chen Y, Li Y, Sun Y, Bai Y, Songyang Z, Ma W, Zhou C, Huang J (2015) CRISPR/Cas9-mediated gene editing in human tripronuclear zygotes. *Protein Cell* **6** (5) :363-72. PMID: 25894090, PMCID: PMC4417674.

11. Braude P, Pickering S, Flinter F, Ogilvie CM. (2002) Preimplantation genetic diagnosis. *Nat Rev Genet* **3** (12) :941-53. PMID: 12459724

Elizabeth McNally M.D., Ph.D.
Elizabeth R. Ward Chair and Director, Center for Genetic Medicine

Elizabeth McNally is a professor in the Departments of Medicine and Biochemistry, Molecular Biology and Genetics at Northwestern University Feinberg School of Medicine where she directs the Center for Genetic Medicine. Dr. McNally is a cardiologist who specializes in caring for inherited forms of heart disease including the cardiovascular complications of neuromuscular disease. Dr. McNally's research has been to identify genes and the mechanisms by which genetic defects lead to heart and muscle disease. A major focus in the laboratory is now to uncover genetic pathways that modify the outcome of neuromuscular disease and its cardiac complications.

Dr. McNally received her undergraduate degree from Barnard College at Columbia University in New York majoring in Biology and Philosophy. She was awarded M.D. and Ph.D. degrees from the Albert Einstein College of Medicine where she participated in the NIH-sponsored Medical Scientist Training Program. Dr. McNally completed training in Internal Medicine and Cardiovascular Medicine at the Brigham and Women's Hospital and Harvard Medical School. Her postdoctoral fellowship was at Children's Hospital in Boston in the Division of Genetics and the Howard Hughes Medical Institute. In addition to her research, Dr. McNally is active physician who established one of the very first Cardiovascular Genetics Clinics in the nation. This clinic, now at Northwestern, provides counseling and cardiovascular care for those with inherited cardiovascular disorders. Dr. McNally advocates for patients and medical research through her work with the Muscular Dystrophy Association and Parent Project Muscular Dystrophy Foundation, where she serves on advisory boards. She was president of the American Society for Clinical Investigation (2011-2012), and a member of the Coalition for Life Science and the Federation of the American Societies for Experimental Biology (FASEB). Dr. McNally has been recognized as an Established Investigator of the American Heart Association and as a Distinguished Clinical Scientist by the Doris Duke Charitable Foundation.

Chairwoman COMSTOCK. Thank you.
And Dr. Khan.

**TESTIMONY OF DR. JEFFREY KAHN,
PROFESSOR OF BIOETHICS AND PUBLIC POLICY;
DEPUTY DIRECTOR FOR POLICY AND ADMINISTRATION,
BERMAN INSTITUTE OF BIOETHICS,
JOHNS HOPKINS UNIVERSITY**

Dr. KAHN. Thank you. Chairwoman Comstock, Chairman Smith, and Ranking Member Lipinski, thank you for the opportunity to testify today on this timely and vitally important subject.

As you heard in the introductions, I am a Professor of Bioethics in Public Policy at the Johns Hopkins Berman Institute of Bioethics in Baltimore.

Also relevant to my comments today, I am also currently Chair of an Institute of Medicine Consensus Study commissioned by the FDA on ethical and social policy considerations of novel techniques for prevention of mitochondrial transmission in women to their offspring. Given that this study is considering issues related to the topic of today's hearing and the work of that committee is ongoing, I will restrict my comments to general observations and an overview of ethical and policy landscape related to gene editing.

I'll focus my comments on three main topics: first, policy history and related areas of science and biomedical research; second, ethical issues raised by gene editing technologies; and third, relevant existing ethical frameworks and approaches to oversight.

The relevant policy history started in 1975, and we heard some mention of this earlier, with the Asilomar Conference on recombinant DNA molecules whose summary statement focused on containment of the risks of creating and working with genetically modified organisms. And with the admonition to avoid experiments that "pose such serious dangers that their performance should not be undertaken at this time," along with a call for continuing reassessment of issues arising in light of new knowledge gained with the experience of the then-new genetic technologies. And we've heard from Dr. Dzau that the National Academies are effectively taking on the same role 40 years later.

These voluntary suggestions that came from the Asilomar summary gave way to more robust oversight as the use of genetic technologies became more refined and with the initial attempts to treat disease in humans. It should be noted that that's 40 years ago, and in the current environment, it would be very difficult for voluntary statements from a collection of even esteemed U.S. scientists to prevent research from going forward internationally, as we've seen already with the publication of the Chinese laboratory experiment.

The ethical issues posed by gene editing and related technologies for modifying human DNA fall into three general categories of concern: first, the implications of the modification of germline DNA, and we've heard some about that already; second, the implications of interfering in processes that should be off-limits to humans. Sometimes those are sort of generally termed "playing God," and that's problematic in some people's minds; and third, the potential for selection or introduction of traits for other than treatment or avoidance of disease, such as physical or behavioral traits or even enhancements.

The focus of much ethical analysis in the application of manipulation of genetic information in humans has been on changes that affect the germline, that is changes that are heritable and therefore able to be passed on to future generations of individuals. The basis of these concerns relate to the uncertainty of the effects of genetic notification, the inability to ''undo'' unintended genetic changes or limit their effects, and the risks of passing on such unintended changes and their consequences to future generations. And that would of course go on forever.

These ethical concerns have been addressed through a range of approaches in order to limit certain types of research or to provide prospective oversight prior to particular proposals being undertaken. We heard from Dr. Dzau that there are restrictions on federal funding of research that involves human embryos. Privately funded research is not affected by these restrictions, though the convention is that research on embryos should take place no later than 14 days after fertilization, and that's a limit also accepted by most countries engaged in research on human embryos. I should also note that there seems to be growing agreement that research should be restricted to nonviable human embryos.

There are a number of institution-level oversight mechanisms that will apply to gene editing. That includes Institutional Biosafety Committees, which are charged with overseeing research with recombinant or synthetic nucleic acid molecules; Institutional Stem Cell Research Oversight Committees, which are charged with, as their name implies, research on human embryonic stem cells and related areas of research. And as the technologies are introduced into human subjects, Institutional Review Boards will be charged with overseeing any potential use on humans.

Lastly, there's a role to be played by the scientific publication community. They have—journal publishers have an increasingly important role to play in setting and enforcing standards of behavior within the scientific community since publication of findings in the peer-reviewed scientific literature signifies the endorsement of the community of researchers.

Journals also play an additional critical role with requirements on the ethics standards being respected, assurances authors' contributions are duly noted, and that human subjects are protected. So there's a role to be played on the part of the journal publication process that will restrict the potential of any unethical research going forward.

Let me conclude by saying the United States has played a leadership role in this area in recombinant DNA and has the opportunity to do so going forward. There will be gaps identified in the process that the National Academies has set out on that need to be identified and addressed, and it is an appropriate time to consider what those ought to be.

Thank you.

[The prepared statement of Dr. Kahn follows:]

Testimony for the Record
Submitted to the
United States House of Representatives Committee on Science, Space, and Technology,
Subcommittee on Research and Technology
for Hearing on
"The Science and Ethics of Genetically Engineered Human DNA"

June 16, 2015

Prof. Jeffrey P. Kahn
Levi Professor of Bioethics and Public Policy
Johns Hopkins Berman Institute of Bioethics
Johns Hopkins University

Chairwoman Comstock and Ranking Member Lipinski, thank you for the opportunity to submit testimony on this timely and vitally important subject.

I am a Professor of Bioethics and Public Policy at the Johns Hopkins Berman Institute of Bioethics in Baltimore. Relevant to my comments today I am also currently chairing an IOM consensus study commissioned by the FDA on Ethical and Social Policy Considerations of Novel Techniques for Prevention of Maternal Transmission of Mitochondrial DNA Diseases. Given that the study is considering related issues to the topic of today's hearing and our work is ongoing, I will restrict my comments to general observations and an overview of the ethical and policy landscape and issues so as not to give the impression of prematurely forecasting the conclusions and recommendations of that committee. I will focus my comments on three main topics: (1) policy history in related areas of science and biomedical research; (2) ethical issues raised by gene editing technologies; and (3) relevant existing ethical frameworks and approaches to oversight.

Related policy history

Starting in the 1970s with the initial discovery and development of recombinant DNA technologies and the ability they brought to manipulate DNA—whether in bacteria, plants, animals, or humans—the scientific community recognized the social and ethical implications of the potential uses of these new technologies.

This began in 1975 with the **Asilomar Conference on Recombinant DNA Molecules**, whose summary statement focused on containment of the risks of creating and working with genetically modified organisms. The summary also identified so-called "experiments to be deferred," which were in its words "feasible experiments which present such serious dangers that their performance should not be undertaken at this time . . ."[1] This admonition was paired with a call for continuing

---

[1] Berg et al., "Summary Statement of the Asilomar Conference on Recombinant DNA Molecules," Proc. Nat. Acad. Sci. 72(6):1981-1984; June 1975.

reassessment of issues arising in light of new knowledge gained with experience with the then-new genetic technology, and included the suggestion of a series of annual workshops, some of which should be at the international level—prescient for today's discussion given the truly global nature of the science. These voluntary suggestions gave way to more robust oversight as use of genetic technologies became more refined and with initial attempts to treat diseases in humans, as I will outline later in my testimony.

Ethical issues raised by gene editing technologies

Scholars and commentators have identified a range of ethical issues posed by gene editing and related technologies for modifying human DNA. They fall into three general categories of concerns: (1) the implications of modification of germline DNA; (2) the implications of interfering in processes that should be off-limits to humans; and (3) the potential for selection or introduction of traits for other than treatment or avoidance of disease, such as physical or behavioral traits or enhancements.

The focus of much ethical analysis in the application of manipulation of genetic information in humans is on changes that affect the germline, that is changes that are heritable and are therefore passed on to future generations of individuals. The basis of these concerns relate to the uncertainty of the effects of genetic modification, the inability to "undo" unintended genetic changes, and the risks of passing on such unintended changes to future generations.

The manipulation of human DNA in such fundamental ways is open to criticisms of interference in processes that we do not sufficiently understand, or that should be beyond human intervention—sometimes noted by the shorthand of "playing God". These concerns reflect the general unease over tampering with the very features that determine our identity, and even our humanity.

With greater precision of the technologies will come the ability to modify genes not only to avoid or cure diseases but to remove or introduce other traits in the interest of enhancement, which some consider a form of eugenics.

How then has the scientific community addressed these concerns?

Existing ethical frameworks and oversight

A range of approaches have been created or promulgated in order to limit certain types of research or to provide prospective oversight prior to particular proposals being undertaken.

Since early stage human gene editing research will require the use of human embryos, existing rules on such research will play an important role. The so-called Dickey-Wicker Amendment prohibits the use of federal funds for research that creates, destroys, or knowingly harms a human embryo.[2] Privately funded research is not affected by these restrictions, though the convention is that research on embryos should take place no later than 14 days after fertilization, a limit also accepted by most countries engaged in research on human embryos.

Institutional oversight

There are a number of institution-level oversight mechanisms that will apply to gene editing research, which apply to various types of proposed research. While there is no single institution-level committee that is currently responsible for gene editing research, one or more may apply depending on the specifics of the research proposed:

Institutional Biosafety Committees (IBCs)—IBCs are charged with oversight of research involving recombinant or synthetic nucleic acid molecules, and review is required for any such research that is "performed at or sponsored by an institution that receives any NIH funding for such research."[3]

Institutional Stem Cell Research Oversight Committees (SCROs)—SCROs are charged with institutional and ethical oversight of research on human embryonic stem cells and related areas of research, following guidelines from the National Academies[4] and relevant state policy where applicable.

While specifics of gene editing research will determine which if any of these existing institutional oversight mechanisms will apply, any research involving human participants must be approved by Institutional Review Boards.

Institutional Review Boards (IRBs)—IRBs prospectively review all research involving humans, requiring appropriate risk-benefit balancing, informed consent of subjects, and monitoring adverse events that occur, in order to protect the rights and interests of those participating in research.

Regulatory oversight

In addition to institutional oversight requirements there are regulatory bodies with roles that are relevant to gene editing research. The NIH Recombinant DNA Advisory Committee (RAC) is charged with making recommendations to the NIH Director "on matters related to (1) the conduct and oversight of research involving recombinant DNA, including the content and implementation of the NIH Guidelines for Research Involving Recombinant DNA Molecules, . . . and (2) other NIH activities pertinent to recombinant DNA technology," and it "makes recommendations on research involving the use of recombinant DNA and on developments in recombinant DNA technology."[5] Critical to the area of RAC recommendations, the NIH Guidelines currently state that "RAC will not at present entertain proposals for germ line alterations . . . Germ line alteration involves a specific attempt to introduce genetic changes into the germ (reproductive) cells of an

---

[2] P.L. 104-99. Thomas H.R.2880. The Library of Congress. Retrieved June 12, 2015.
[3] NIH Guidelines for Research Involving Recombinant or Synthetic Nucleic Acid Molecules (NIH Guidelines), Nov. 2013.
[4] Guidelines for Human Embryonic Stem Cell Research, Washington, DC, National Academies Press, 2005.
[5] Charter, NIH Recombinant DNA Advisory Committee, June 30, 2013.

individual, with the aim of changing the set of genes passed on to the individual's offspring."[6] This indicates a current effective prohibition on the use of germline modifying technologies for areas of research within the purview of the RAC.

Should there be application for an Investigational New Drug, FDA review and approval would be required prior to the administration in humans, a process that in the case of gene transfer takes place in parallel with and informed by the review process of the RAC.

The role of scientific journals

Lastly, scientific journal publishers have an increasingly important role to play in setting and enforcing standards of behavior within the scientific community. The goal of credible researchers worldwide is publication of their work in the peer-reviewed scientific literature, which signifies the endorsement of the community of researchers working in similar areas and acts as a means of sharing advances in ways that credits the researchers and labs that achieve them as well as moving the field forward. But journals play an additional critical role with requirements that ethics standards are respected, through assurances by authors regarding their contributions, that human subjects protection standards are met, and that conflicts of interest are disclosed and sufficiently addressed.[7] Journals could play a similar role in relation to the publication of research involving gene editing technologies in humans. The study recently published in *Protein & Cell* was reportedly rejected by both *Science* and *Nature* prior to its acceptance and eventual publication, with some reportage indicating that ethics reviews played a role in the decision to reject it. There has been at least one editorial suggesting that journals could add to existing requirements and require that authors provide the details of the ethics review of any gene editing-related research as a condition for consideration of publication.[8]

Conclusion

The United States has long played a leadership role in both science and in the responsible use of the advances discovered or developed. This was certainly the case with the introduction of recombinant DNA technologies. It is critical that we continue to do so as the new and powerful genetic technologies we are discussing today become both more precise and more widely available. Existing oversight approaches may provide part of what can be the framework for addressing many of the issues raised by gene editing technologies. However, there are likely to be gaps in oversight of applications of these technologies, and work must be done to (1) identify these gaps in both in the near and longer terms, and (2) craft appropriate guidelines to bridge these gaps in order to most

---

[6] NIH Guidelines, Nov. 2012, Appendix M.

[7] International Committee of Medical Journal Editors, "Recommendations for the Conduct, Reporting, Editing, and Publication of Scholarly Work in Medical Journals," December 2014; http://www.icmje.org/recommendations/.

[8] A. Sharma, C.T. Scott, "The Ethics of Publishing Human Germline Research," Nature Biotechnology 33(6):590-592; June 2015.

appropriately and successfully address the ethics and oversight issues arising as the use of these technologies expands and becomes more sophisticated. This work must reflect input and contributions from the scientific community, ethics experts, policy makers, and a range of public stakeholders. Only then will we achieve a robust and credible policy framework that will assure the responsible use of these technologies while achieving their promise for advancing scientific knowledge and human health.

Thank you.

Jeffrey Kahn is the Robert Henry Levi and Ryda Hecht Levi Professor of Bioethics and Public Policy in the Johns Hopkins Berman Institute of Bioethics. He is also Professor in the Department of Health Policy and Management in the Johns Hopkins University Bloomberg School of Public Health. Prior to joining the faculty at Johns Hopkins, Prof. Kahn was Director and Maas Family Endowed Chair in Bioethics in the University of Minnesota Center for Bioethics. His research interests include the ethics of research, ethics and public health, and ethics and emerging biomedical technologies; he speaks widely both in the U.S. and abroad, and has published four books and over 125 articles in the bioethics and medical literature. He is an elected Fellow of the Hastings Center, and has chaired or served on committees and panels for the National Institutes of Health, the Centers for Disease Control, and the Institute of Medicine.

Chairwoman COMSTOCK. I thank the witnesses for their very interesting testimony.

And now I remind Members that Committee rules limit questioning for five minutes.

And the Chair—as Chair, I recognize myself for five minutes.

So I wanted to hear a little bit more, Dr. Doudna, from what you mentioned where you had initiated the discussion with the scientists in California and bioethicists and others to address this. Some of the key—and maybe expand a little bit more on some of the key issues you found and how we avoid maybe where China might go and how we do define those boundaries and kind of what some of the concerns were that were raised. And if anyone else would like to address that, too, but I thought I'd start with you and maybe expounding a little bit more on this group.

Dr. DOUDNA. Sure, thank you.

I think, you know, what was interesting to me in that conversation was that it was a wide-ranging discussion that started with sort of the—maybe the way that we've been discussing this technology so far here in the Committee and eventually got to a point where, as somebody around the table said, you know, there may come a time when we would consider it unethical not to do editing in the germline for certain kinds of applications such as some of the things that Dr. McNally mentioned.

So I think that it's very important to appreciate that, you know, this is a powerful technology that is—you know, we're sort of looking at it here from the perspective of safety and ethical concerns but I think they're also—you know, that can be turned around. And that was something that came out of our discussion in California that I found very interesting.

And the other thing that we discussed was the fact that, you know, this technology, unlike previous technologies for genome editing, is very simple to employ relatively. I mean it's something that, you know, people that have expertise in molecular biology can fairly readily use in their laboratories. So I think, you know, the reality is that it would be very hard to really put regulations on this in terms of research applications.

And I think that means that we just have to be thoughtful about providing leadership in terms of the—you know, the—as Dr. Kahn was saying, in terms of the way papers are published and reviewed among scientists so that, you know, the scientific community helps to provide the kind of direction and, you know, vision for the way this should move forward that hopefully will be respected by many others. But I think the reality is that it—you know, it is a technology that's just very widely available now and is being employed worldwide.

Dr. DZAU. I'd like to echo what Dr. Doudna said. I think we at the Academies feel very privileged to be asked by the scientists—Ralph Cicerone, the President of National Academy of Science, myself, got lots of phone calls and emails to say, you know, the Academy really should take this on because we have done so in the past, as you heard, during the time of recombinant DNA, Asilomar conference, the human cloning, and also the whole issue of stem cell guidelines, as well as mapping the human genome. So we have to do this.

I think that in the context of what's being discussed, it's really important that we have people from all different disciplines, not only scientists but ethicists, legal experts, medical physicians, and others all engage in this discussion because at the end of the day what we want to do is to find what appears to be the consensus of what the right thing to do is for our country and so we look forward to working on this.

Dr. KAHN. So I would just like to echo what you've heard from my colleagues on the panel and add a little bit to expand on what I said about publication. It's difficult for the scientific community to identify any one thing that it can take on in an international way, but scientific publication knows no borders. And there have been calls now for—before publication of any gene editing research that there be a disclosure of the ethics review that that research underwent. That did not happen so far as we know with the Chinese research that was recently published, and in fact, the reports have been about Science and Nature rejected that paper when they considered it and at least in part on ethics grounds. So that would be one way to create some—a framing of what would be acceptable as determined by the scientific community itself through the peer review process.

Chairwoman COMSTOCK. Thank you. And, Dr. McNally, did you have anything to add? I have a few more seconds before I turn——

Dr. McNALLY. I would also remember that through this process it's not just the scientists and the physicians and the regulatory bodies but to also remember the patient advocacy groups and the patients themselves. They're going to be the loudest voice in this process and we need to hear what they have to say.

Chairwoman COMSTOCK. Thank you. I appreciate it.

Now I recognize Mr. Lipinski for five minutes.

Mr. LIPINSKI. Thank you.

Before I follow up, I just want to make sure we're more clear on this when we get into the subject.

Dr. Doudna, you and the group that have met, you've called for a temporary moratorium on the use of the technologies on human embryos. Is that correct? And what led you to that and are there particular milestones and discussions of the science that you're working towards and then see ending the moratorium? And I want to get the other panelists to comment on that.

Dr. DOUDNA. So I would say what led us to that initial meeting in California was the appreciation that this technology would likely, you know, be functional in the human germline, and furthermore, that it was possible that people could do this fairly easily and perhaps working in jurisdictions where there would not be regulatory oversight of such experiments. And what was interesting was that at that meeting we actually heard about the work that was subsequently published for the Chinese group, so it became apparent that, you know, the subject of that conversation was very timely.

I think that, you know, going forward we really have to appreciate the difficulty in putting in place regulations that will be, you know, followed internationally. On the other hand, providing oversight and leadership in—by respected scientists in the United States and with our international partners I think will be a very important thing to do and that's really what we wanted to achieve at that meeting was how to proceed with that sort of approach. And I think now having the National Academies involved in organizing larger meetings around this issue is a very desirable outcome.

Mr. LIPINSKI. Does anyone else want to comment on a temporary moratorium on human embryos?

Dr. DZAU. So the National Academies' position is that we need to be thoughtful, comprehensive, scientifically driven, and independent so we are convening those bodies and obviously supporting the idea that we should be very cautious. We have not actually taken the position per se because until our work is done, you know, we would be looking at what the most objective way to approach this when our work is done.

So I think some of these meetings are critically important and important in the sense that we need international scientists to be involved with this conversation as well. We have in fact on our advisory board scientists from the Royal Society in London, scientists from the Chinese Academy of Science involved with this conversation because it's really critical to consider all aspects.

As Dr. McNally pointed out, you know, if you're using certain conditions, I think we have no doubt that this would be enormously helpful to patients and to mankind. The question is what are the areas we need to be concerned about?

That being said, I think technology is still early because even when you look at the Chinese publication, they were looking at incomplete editing and many other issues that research has to go forward to understand the safety, the off-target effects, and the efficacy before human application. So that's our position.

Mr. LIPINSKI. Okay. Dr. McNally?

Dr. McNALLY. I would like to echo what Dr. Dzau just said which is that I appreciate very much the leadership by Dr. Doudna and the group of scientists that got together but I think the IOM convening an international effort to really look at what the possibilities are I would think it's reasonable to wait until we learn what that outcome is. It's important to have a careful and to have an international look at what the best recommendations really are at this point in time.

Mr. LIPINSKI. Dr. Khan?

Dr. KAHN. Just very briefly, the kind of work people are talking about on an international level is really critical in establishing principles and guidelines for how to move forward. I think moratoria—blanket moratoria are probably not what we need but figuring out when we can go forward and how and under what kind of restrictions. So certainly in vitro only and probably, you know, for a very long time until it's determined that the milestones have been met in order to move forward into humans. So that's the good work that entities like the Academies can engage in and I think critical at the international level.

Mr. LIPINSKI. Thank you all very much. It's—yeah, I think there are very difficult ethical questions we need to deal with here and I know myself I certainly don't know enough about the technologies, some of the specifics there, and I think that makes—certainly makes a difference. And then we do get to a question that

Dr. McNally said there are things that we certainly want to cure and then the question always is how far do we take this? But we're certainly not going to solve that here so thank you very much.

Chairwoman COMSTOCK. Thank you.

Now I recognize Mr. Moolenaar for five minutes.

Mr. MOOLENAAR. Thank you, Madam Chair.

And thank you for your presentations today and my compliments to you on the work you're doing. It's very sophisticated and it seems that it has tremendous potential for good.

On the area of ethics of this I wanted to just explore a few different areas. Dr. Khan, I wanted to start with you. When you're advising on the ethics of a new genetic technology, are there certain ethical lines that should never be crossed, and if so, how are those lines drawn?

Dr. KAHN. That's a great question and how much time do we have I guess is going to be part of my answer. But of course those of the first kind of questions that we ask.

I will say there has been a long-standing line that has not been crossed and that is modification of the human germline. That has held for decades since the beginning of recombinant DNA technology was available and begun to be implemented.

One thing that none of us have actually talked about is the recombinant DNA Advisory Committee, which is a committee of the advisory to the Director of the NIH, which has in its guidelines a statement that they will not consider any proposed research that modifies the human germline. So it's effectively a prohibition from that research going forward. Now, it only applies to research that is subject to the oversight of that committee, but that has been a bright line.

The other thing which we sort of skirted around but haven't maybe set explicitly is that when there's such uncertainty about the risk and the outcomes, we go slowly. So that's the kind of a soft answer to your question, a mushy one, but I think that's an important principle to—just to articulate in a way that's explicit. When we know it's time to go forward is a harder question but we always start slowly, especially when we're talking about potentials for modifying humans.

Mr. MOOLENAAR. Thank you.

And then, Dr. Doudna, you're one of the new developers of this new technology, and what was it that first sparked your concern over the ethics of its use, whether it was in China or when did you start being concerned about that?

Dr. DOUDNA. I guess I realized the potential for this technology to operate in the germline first when scientists began to do experiments of that nature in animal models of disease, including mice and rats, and then it really came home to me in—it was last year—may be almost a year-and-a-half ago that a group again from China published a paper in which they had modified the germline of monkeys and made genetically modified monkeys. And that actual monkey model is used very commonly for studying human disease and so it seemed very likely at that point that there was no reason to think the technology wouldn't also work in the human germline.

So, you know, I think we've seen now in the scientific community that this technology is very democratic in the sense that it works across many different types of cells. It doesn't seem to be limited to a particular system.

Mr. MOOLENAAR. And what role does—you know, I guess one of the things that occur to me is the whole area of consent. What role does that play in this process when someone is giving consent or not?

Dr. DOUDNA. Are you directing that question to me?

Mr. MOOLENAAR. Yeah—actually, all four.

Dr. DOUDNA. You know, I would—maybe I would defer that to Dr. McNally.

Dr. MCNALLY. Well, as a member of an IRB for about 15 years, that's actually—you know, right now, consent in a research study in that case would be the parent; the mother would be the person providing consent. If a study were to go forward right now, that's who would be providing consent because it would be her materials that were being used for that purpose. So there isn't—from the standpoint of human subjects strictly talking like Institutional Review Boards, a fertilized egg is not an individual that provides consent and also even—you know, as a—if it were a child, a parent provides consent for that if that answers your question.

Mr. MOOLENAAR. Yeah. Thank you.

Dr. KAHN. So it's—I would sort of ask consent from whom and for what? So you've heard a version of the answer to that question. If we're talking about the donor of the materials that would be researched upon, that's one set of questions. If we're talking about modifying an embryo that might one day be implanted into a woman's body and developed into a child where we have a very difficult conceptual problem. How do we think about consent on behalf of somebody who's not yet been born? And so those are the really interesting ethical questions that we will need to confront. We're certainly nowhere near thinking about doing that kind of application, but those are the kinds of questions that need to be identified, articulated, and addressed in efforts that are like the ones the Academies are taking on.

Mr. MOOLENAAR. I would agree with that. Thank you.

Chairwoman COMSTOCK. Thank you.

And now I recognize Mr. Tonko for five minutes.

Mr. TONKO. Thank you, Madam Chair.

This hearing shines a light on a difficult but indeed important issue and I appreciate my colleagues' focus on the ethical and legal issues surrounding this new technology.

I'm also assured to see that experts in the medical and scientific community are coming together to debate this issue and to discuss potential policy implications. However, as we explore the boundaries of what science is capable of and what is ethical and what should be legal, we should also take a moment to appreciate just how remarkable these advances are. We recognize that these new gene editing technologies, including CRISPR, are the outgrowth of decades of fundamental research supported by federal agencies, including the National Science Foundation.

So to Dr. Doudna and Dr. McNally, could you please speak to the importance of federal investments in fundamental research, especially the need to support research that may not have any known commercial application at the time?

Dr. MCNALLY. Again, how much time do we have?

Mr. TONKO. Well, the Reader's Digest version.

Dr. MCNALLY. I think everybody sitting here at this table can say it cannot be overstated how important the federal investment is for research. There are many people who would love to see a lot of research funded in the private sector but there are certain aspects of particularly fundamental research that will never be covered in the private sector. Sequencing the genome is a great example of that.

And we can't move forward without that federal investment and I think all of us here would say, you know, it's been fairly tough times in the last few years with what's happened with budgets and what's happened with research and watching the effect that that has had on the scientific community here in the United States where we have actually seen the size and shape of the scientific community shrink in the last few years, especially when we look across the world and we see it growing elsewhere. So, yes, federally funded research is absolutely essential for these types of basic observations.

Dr. DOUDNA. Right, so I echo everything that my colleague Dr. McNally just stated and I want to also add that, you know, there's a tremendous opportunity for the United States to invest in basic science. I mean I think traditionally our country is been a leader in science partly because we have invested in science that was, you know, curiosity-driven research. It was not necessarily targeted on curing certain diseases, and I think that we've seen again and again, especially I would say with regard to technologies, they tend to come from unexpected types of projects such as the CRISPR system is a great example of that but there are many others. And I think also it's important to appreciate that commercially, you know, these things then have big implications in terms of companies being able to take over and, you know, develop technologies that are discovered in academic laboratories but then apply them in all sorts of different ways.

Mr. TONKO. Dr. Dzau, I think you wanted to respond to that, too?

Dr. DZAU. Well, I totally agree with my colleagues and they're particularly emphasizing support for basic research because if you think about this technology, it was done on bacteria, and without thinking of application human and look where it is today. And we can count so many important breakthroughs that come this way. So the ability to support fundamental basic research is critically important.

Also, I think along the issues about saving human lives, creating jobs, is our global competitiveness situation, we are truly concerned that we don't continue this level of investment that the United States becomes less competitive than many other countries which are investing heavily into basic and translational research.

Mr. TONKO. Thank you.

Dr. McNally, in your testimony you mentioned how these technologies may be able to be used in somatic or mature cells to treat and potentially cure diseases such as sickle cell anemia and muscular dystrophy. Can you please elaborate on this possibility and what is that range of therapies and cures that we might only imagine?

Dr. MCNALLY. I think it's widely anticipated that sickle cell will be one of the first things that's cured by this where you could take a cell out because it's a bone marrow cell. You could correct it with CRISPR/Cas and return that person's bone marrow cells so it's not a transplant situation; you're returning their own cells to them. And that's ongoing right now and I anticipate that we will see that.

For my field I work in muscle diseases. Duchenne muscular dystrophy is a very challenging, challenging area and right now we're looking at technologies where we're taking small antisense compounds where we would have to treat that individual for a lifetime every day with those compounds. Again, if we could get the cells, correct them, and reinsert them back in, that would be a one-time treatment and a lifetime treatment for that individual. So I think we're seeing a few examples where it's definitely heading that direction.

Mr. TONKO. Thank you so much.

Madam Chair, I yield back.

Chairwoman COMSTOCK. Thank you.

I now recognize Mr. Palmer for five minutes.

Mr. PALMER. Thank you, Madam Chair.

Several folks have mentioned the research on stem cell. Dr. Tim Townes at UAB is a very good friend of mind and doing world-class research in that area.

I've got just a few questions. Dr. Dzau and Dr. Kahn, the United States and Europe have often disagreed on regulations and policies and other areas of biotechnology, for example, genetically modified organisms. How does that impact international scientific cooperation? And do you anticipate a similar challenge in human gene editing?

Dr. DZAU. Well, the intention is that when you get scientists, regulators, ethicists together from different countries, I think responsible individuals would begin to talk about what would be responsible behavior. That I believe is the starting point. And I do agree with you that we have some differences in the regulation. But I think overall in the issue we're talking about today, which is the application in germline gene editing, I have a sneaking suspicion—although I don't want to fully predict this until the work is done—there's great agreement about the concern about creating successive generations of individuals whose genes have been altered. So I have a feeling that we actually will get agreement if not harmonization of many of these thoughts, and certainly it's our hope that we will reach that in our initiative.

Mr. PALMER. There are current regulations that prohibit federal funding for human—for research on human embryos and the FDA requires—must issue an investigational new drug application before a biological product may be used in humans. Do you think these kind of safeguards are adequate to prevent the kind of experiments that we're concerned about?

Dr. DZAU. I'd like Jeff to answer this as well because, as you heard, he's in the midst of leading one of our initiatives on mitochondria DNA replacement.

Mr. PALMER. Um-hum.

Dr. DZAU. But I think we very much look into what is the right regulatory framework particularly for issues like this, and I would say that, you know—and because he would have an opinion but my feeling is that this work needs to be done to get better clarity about when and how we would regulate some of these areas.

Mr. PALMER. Dr. Kahn?

Dr. KAHN. Thank you. In answer to your question, I think that the FDA will play a critical role although it will only play its role towards the end of the story so the very basic research will be done prior to anybody thinking about an IND application or introducing it into a therapeutic context. And so we need to think about the entire translational pipeline as it were and all of the issues that will arise along the way and make sure that we have appropriate oversight for each of them.

The FDA is clearly thinking about the issues that you are and are trying to figure out how their framework ought to apply.

So first—for first in human applications effectively, and then once something is licensed, how to control its use and dispersal. So one of the things that we all worry about is the so-called off-label use of the new technology. So even though the FDA may approve it only for a limited purpose, once it's licensed, it's hard to control. The FDA has some new tools that may actually make them more feasible to do and I think as we get closer to the kinds of technology were discussing entering the therapeutic marketplace that there will be stronger safeguards in place.

Mr. PALMER. Well, with these safeguards—and staying on that line of thought—is there any worry that if the United States doesn't use this research that we could fall behind our international competition and, you know, could new regulations of this technology in the United States put our researchers at a disadvantage or cause them to move their research overseas, Dr. Kahn?

Dr. KAHN. I'll let Dr. Dzau and others speak to this, too, but there always is talk about that and of course it became a bigger issue when the stem cell research—human embryonic stem cell research field really began to grow. And there are now international stem cell research societies which try to address issues that are clearly not governed by borders. And what we don't want of course is to have people who are doing the best science in the world think about leaving this country because it's easier to do elsewhere. So we need appropriate controls but not those that squelch the science. Finding that sweet spot of course is the challenge.

Mr. PALMER. One last question and you can follow up on that as you answer this also, but do you believe that we'll be able to get China and the United Kingdom and these other nations to work together to influence change and Europe to adopt similar standards—similar safeguards?

Dr. KAHN. Maybe Dr. Dzau can speak to that if that's okay.

Dr. DZAU. Well, we're starting with getting the major science academies involved so we have the National Academy of Science and Medicine in the United States; we have the Chinese Academy of Science where you would think that they have tremendous influence on the way the conduct of science is being carried out. We have the Royal Society and we intend to include many international bodies.

So I think the starting point clearly is with the scientists saying what would be the right thing to do. One can imagine that we may have to escalate this conversation further depending on our findings.

I just want to point out the question that you asked, which is the regulation aspect of this. In fact, you know, we do need a lot more clarity in this country. We haven't talked about the use of gene editing in nonhuman cells, plants, insects, and those changes are—that science is really ongoing. We are producing possibly new species that would turn around much faster because of a much shorter cycle time, reproductive time. So that regulation also has to come in to say what in fact is considered safe and what's considered as environmentally sound. And in fact the National Academies is also looking at a study looking at this issue. And as you know, in this country the USDA and FDA are involved with animal and plant regulations. So we're also trying to give the right recommendation to fortify our regulatory processes.

The final question you asked earlier about what the right thing to do is I think from our perspective at the National Academies I think our first issue is human protection and doing the right thing for society. I understand of course the potential loss of scientists, et cetera, but I do think that we have to actually take the high road to say what's right for us first. And I have a feeling everything else would follow and fall in the right place.

Mr. PALMER. Thank you.

Thank you, Madam Chair.

Chairwoman COMSTOCK. And I now recognize Mr. Swalwell for five minutes.

Mr. SWALWELL. Thank you, Madam Chair, and thank you to our panelists.

My first question relates to something that a number of the Members have brought up, which is it seems that America and our investments in federally funded research have been in decline, and as a result, our successes have been in decline and our ability to attract and recruit and retain some of the best and brightest scientists may be in decline. So let's just for argument's sake go back to 1995. I was 14 years old. And the Human Genome Project was just starting to get off the ground and many great results came out of that. That was 20 years ago. Would each witness just say more or less has America and its—have you seen the investments that we've made as far as federally funded research, has that made America more or less exceptional in this field? So just tell me more or less. And I'll start with Dr. Dzau, just one word.

Dr. DZAU. More.

Dr. DOUDNA. More.

Dr. MCNALLY. More.

Dr. KAHN. More.

Mr. SWALWELL. So I'm confused because each of you has said that our investments have been on the decline and so you're telling me that we actually have made more investments since 1995 and you believe we are more exceptional now in these fields?

Dr. MCNALLY. You picked 1995.

Mr. SWALWELL. But comparing——

Dr. MCNALLY. If you said 1995——

Mr. SWALWELL. Sure.

Dr. MCNALLY. —to 2005, we would say more. If you would say 2005 to 2015 we would all say less.

Mr. SWALWELL. Is that right, less, Dr. Dzau?

Dr. DZAU. [Nonverbal response. Nodded in the affirmative.]

Mr. SWALWELL. Dr. Doudna?

Dr. DOUDNA. [Nonverbal response. Nodded in the affirmative.]

Mr. SWALWELL. Dr. McNally?

Dr. MCNALLY. [Nonverbal response. Nodded in the affirmative.]

Mr. SWALWELL. Dr. Kahn?

Dr. KAHN. [Nonverbal response. Nodded in the affirmative.]

Mr. SWALWELL. Okay. So that was my question. You would agree that we have become less exceptional in the field of genetic engineering as far as it relates to human DNA since 1995, that we've been on the decline?

Dr. MCNALLY. 2005.

Mr. SWALWELL. 2005 is the——

Dr. MCNALLY. Yes.

Mr. SWALWELL. —point, Dr. McNally?

Dr. MCNALLY. Yeah.

Mr. SWALWELL. So what's exciting about this research and this field is the potential for us to conquer diseases before they conquer us. And I look at the example of Huntington's disease, which affects anywhere from 30,000 to 200,000 people, and it's a disease that is so cruel it steals your memory and affects your muscular system.

And I'm just wondering, maybe if Dr. Doudna can tell us and have others weigh in, what can the United States do specifically to take leadership in this area if we have the appropriate funding so that we can conquer these diseases?

Dr. DOUDNA. So I think I know, what—again sort of maybe echoing something that I mentioned earlier and that has been discussed here, I think, you know, the United States has been a real leader in basic research for a while and all of us are concerned that we see that edge slipping away over time. And so I think that, you know, the investment in fundamental research that will allow scientists to understand, for example, genome engineering technology like we're discussing today, how does it operate, how can we deliver it to patients, how do we ensure that it's operating as we intend and not creating unintended consequences, that it's safe, that it's effective.

All of those lines of research are going to require, I would say, a combination of efforts by people like me that do basic research and people like my colleagues who are medical doctors and think about clinical issues. We need to be putting our efforts together and that has to be I think supported by federal funds.

Mr. SWALWELL. And how can we tell the story to the American people who look to us as a Congress to make the decisions when it comes to funding with so many competing priorities? How can we tell them that something—like $1 invested in basic research where you may not be able to tell us what disease you're going to be able to cure 10 to 15 years from now but there's still—the taxpayer is looking to us to, you know, hold accountable the funding. Like how can we better tell the stories of the science community about what we could see from this down the road and how we could truly, you know, attack and cure some of these diseases? I think that's probably one of the biggest challenges. And maybe Dr. McNally would want to answer.

Dr. MCNALLY. Yeah, I mean we've seen incredible advances that have come out of basic research, and I'll talk about my field, which is cardiology. It was out of basic research of understanding the LDL receptor that led to statins, the drugs that probably a lot of people in this room take, and we've seen a direct translation to a reduction in the rate of heart attacks. I mean it's a very different world than what it was when I first heard of my training 20 years ago. We don't see acute heart attacks like we used to and that came as an outcome of basic research not that many years ago.

So we can tell the same story for heart disease. We can tell the same story for many cancers. Cancer and heart disease are the major things that kill people so we've made huge headway in that.

Mr. SWALWELL. Great, thank you. And, you know, I know every Member up here has thousands of people in their district who go to bed on their knees praying that people like you will make discoveries that will make them or their relatives live healthier lives. So thank you for what you're doing and hopefully we can do the right thing here and better fund your initiatives.

And I yield back.

Chairwoman COMSTOCK. Thank you.

And I now recognize Mr. Westerman for five minutes.

Mr. WESTERMAN. Thank you, Madam Chair, and thank you, panel, for your invigorating testimony.

Dr. Dzau, what's your anticipated timeline for the initiative on human gene editing?

Dr. DZAU. We are convening an international summit in the fall—late fall and that should be a—I think really an important meeting that will get together international scientists. As you heard from our previous experience at Asilomar in the 1970s, you know, one would engage in the discussions that Dr. Doudna and McNally and Kahn put forth, and hopefully the findings of that meeting will be published very shortly after that.

Perhaps equally important is a concurrent deep-dive study which we're conducting that would involve analysis, research, the assessment of risk/benefits, as well as a regulatory framework and ethical issues, and that usually—what we call a consensus report would take as long as about a year, although we're hoping that we would try to move it faster for exactly the reasons that we've been talking about. Such a report with very specific recommendations which we're willing to put forward under both public hearing and also closed discussions will be available to you, to the Nation, and to many others. So I could say the time frame is late fall to sometime next year but hopefully as early as we possibly can to put out that report.

Mr. WESTERMAN. So how great are your concerns that the initiatives may not be able to keep up with the breakneck speed of the technology as it moves forward?

Dr. DZAU. This is exactly why we need a summit, a meeting first, where key scientists—and we're going to engage a large number—will discuss about what would be considered as good conduct, good oversight, understanding risk, et cetera, so that the scientific community which is really driving most of this research understands those issues and have in general agreement.

We also believe of course in our study itself that becomes a definitive document by which you and others can be informed to say what are the right decisions based on consensus.

Mr. WESTERMAN. So you said the scientific community is driving the research. As the National Academies' work on the new initiatives for building the framework, do you believe the scientific community will embrace and voluntarily follow the guidelines or will there need to be regulations or laws put in place?

Dr. DZAU. Well, you know, as you already heard, I think there are many responsible scientists, particularly the ones who are leading this field, who feel already that we should, you know, put a moratorium and not slow it down. So I do believe that already going into this meeting, although there may be many different opinions, there's probably general agreement that we need to slow down this area until we have a much clearer point of view about where we should be going and the clarity in terms of regulation.

The problem I'm concerned about is that we rush into this too fast. We don't really want to and it's such an important issue. We've got to be very thoughtful. That being said, we understand the time urgency of the issue—situation.

Mr. WESTERMAN. All right. And the advisory group to the initiative that was named yesterday, it includes scientists and researchers from China and the United Kingdom. Do you believe these participants will help influence Asia and Europe to adopt similar standards?

Dr. DZAU. We certainly hope so, and in fact I was on the advisory group. Our intention is that in an international summit we include a lot more scientists from Asia and every part of the world, Europe, et cetera, to be part of this discussion.

You know, it's interesting when we think back on U.S.'s position. You know, when the Asilomar Conference come about, the United States was the main show in town about technology, so among the U.S. scientists, you can imagine there's agreements, you know, so there's a general way of saying, you know, what do we do next? Here, we really need to include international scientists.

Mr. WESTERMAN. So what capacity and infrastructure does the Chinese Government have in place for regulating human scientific research?

Dr. DZAU. You know, I'm not that familiar with the regulation at this point in China. As you heard from my colleague, Dr. Kahn, there is some speculation over what's there and what's not there. I think we would hope that that meeting will bring out with clarity what each country's position is, what's the regulatory position, what's the scientific position so that we can all come together and examine this together.

Mr. WESTERMAN. Thank you, and I believe I'm out of time, Madam Chair.

Chairwoman COMSTOCK. All right. Thank you.

And I now recognize Mr. Foster.

Mr. FOSTER. Thank you, Madam Chair.

In addition to the international summit that you're having this fall it strikes me that a full-blown National Academies study may have merit to do a deeper dive into this. And so, Dr. Dzau, would a letter signed by Members of Congress, for example, help you in recruiting assistance for this sort of effort?

Dr. DZAU. I think it would be outstanding in fact if we got a letter from Congress on this. We took this upon ourselves as a National Academy because that's what we do, and we know it's the right thing to do. We have the support of scientists and others to say go forward. But I think it would be tremendously impactful if Congress would provide us with that kind of support a mandate to go forward with this.

Mr. FOSTER. Okay. Just a quick question, today what's the rough cost and time to get a mouse model with a specific genetic modification? Does anyone—just roughly within a factor of two—you know, is it $10,000 or $5,000 or——

Dr. MCNALLY. Yeah. Well, with CRISPR/Cas initiated, yeah, you'll probably decrease the cost in half and decrease the time in half. So if it was $25,000, it's probably closer to 10 now, and if it was a year, it's probably closer to 6 months. That's counting breeding time.

Mr. FOSTER. Well, when—so one of the things slowing down the application of this is a lot of—sort of two classes of worries about potential dangers. The first is so-called off-target effects where, in addition to the genetic modification you want, you get inadvertent modifications to the genome. And it's my understanding that the technology is evolving rapidly and that—I was just wondering if you—I'll ask you to go out on a limb I guess, and Dr. Doudna first, about—you know, if you just look at the rate of progress on this, what is the rough timescale where we can expect—where we might expect you'll be in a position that it could be used, you know, "safely" on humans?

Dr. DOUDNA. Well, I think one has to, you know, sort of distinguish what types of applications we're talking about. I think if we're considering application to treat a disease like sickle cell anemia where the editing could be done on cells that are taken out of the patient and then validated before they are reintroduced into the body, I feel that that is likely to happen, you know, the next year or two honestly. I think it will be very——

Mr. FOSTER. Right, so no technological development there?

Dr. DOUDNA. No, because I think we already have the ability to, you know, validate the correct sequencing—correct editing was done by DNA sequencing in that sort of a scenario. I think if you're talking about an application like, you know, we want to introduce the tool into a patient's body and where you want editing to occur in the body, then we're—that's further off. First of all, we don't have the—very good ways to introduce this into specific tissue types yet, and also we don't have good ways to validate that the correct type of editing was done without off-target effects, as you implied. But I think for any kinds of applications where we can do the validation outside of the body, that's going to move forward in the next year or two.

Mr. FOSTER. Okay. And then if I raise the stakes further to germline editing, is that something that may just never happen? It may never be reliable enough? Or is it a reasonable guess that within the next five years that you'll be able to validate the germline editing, that it has taken place correctly and with high enough confidence that—

Dr. DOUDNA. Well, I'm very interested to hear my colleagues' answer to that question but I guess my answer would be that it will depend on the way that research is enabled around, you know, that sort of application. I think if it's possible to do experiments in germ cells so that we can understand how this technology works, operates in those types of cells, then I think, you know—boy, it's always hard to put a timeline on things but, you know, certainly within a few years it'll probably be to the point where one could, you know, employ it for that sort of application. But I don't know if you—my colleagues would agree with that or not.

Dr. MCNALLY. I agree. I think that's a reasonable timeline, five, ten years if you had to guess.

Mr. FOSTER. Yeah. I'm trying to, you know, get some idea——

Dr. MCNALLY. Yeah.

Mr. FOSTER. —of what the response time from Congress and our society has to be for that.

There's a second class of potential dangers having to do with just misunderstandings about what the effects of a specific genetic change will be on the characteristics of the adult organism. And, you know, over the spread of, you know, different things from simple conditions like sickle cell to complex things like, you know, personality, you know, what is your guess for the timescale that we're looking at there from right now to never?

Dr. DOUDNA. My answer is certainly much longer. I think we— I think—and that—to me that's not limited by the genome editing technology as much as it's limited by our knowledge of the human genome.

Mr. FOSTER. Okay. Thank you. My timer has gone down. I yield back.

Mr. MOOLENAAR. [Presiding] Thank you. I now recognize Dr. Abraham.

Mr. ABRAHAM. Well, after I read you all's testimony last night I went on and read each of you all's bios. Your parents must be very proud.

We all in this room I think understand the potential that this type of research can lead to not only in the human endeavors but in plant technology, curing world hunger. So it's applicable to so many aspects of humanity.

And to—I think it was Dr.—your point, Dr. McNally, that it probably would be unethical if we have a child with sickle cell in the ALL and this therapy is available not to offer it.

Dr. Doudna, I salute your intelligence for recognizing it even though I know you weren't particularly looking for the CRISPR/Cas9 technology that you recognized it. I compared it last night— I was reading—to Fleming discovering penicillin. He had been—I think maybe this CRISPR can save as millions of lives as penicillin has, so kudos to you guys for what you do.

I guess the question I'm leading up to, the old adage if you get two doctors in a room, you get three different opinions, as we know very true. That's certainly on my end of the stick.

Dr. Doudna, you published I think an article in a science magazine that you wanted this moratorium and, Dr. Dzau, you've been pushing for. Are you guys getting some pushback from members of your community that says, you know, no, we don't need a moratorium and let this thing go? Let this genie out of the bottle and don't put it back in?

Dr. DOUDNA. Well, I can tell you what I'm seeing. I think that, you know, at around the same time that we published the perspective in Science magazine a related perspective was published in Nature magazine from a different group that actually called for I would say real moratorium even on research. So that group basically was advocating not proceeding with any kind of research on human germ cells using genome editing technology. And I just want to point out that in the group that I met with in January, we actually discussed that and felt that actually we—in our opinion research on those types of cells, appropriately regulated, should be enabled, just not clinical application.

Mr. ABRAHAM. Dr. Dzau, when—Dr. Kahn, when you had the— I think it was back in 1975 I was reading last night y'all had a recombinant DNA moratorium that you tried to put forth voluntarily. In the international community, was that followed? Has that been pretty much adhered to throughout the last few generations?

Dr. KAHN. So a little bit before my time but it's a landmark in the area of science policy and then it was a voluntary moratorium instituted by the scientific community. The truth is we haven't seen anything like that since. It was a very important undertaking. And I think when the scientific community got together to talk about the implications of recombinant DNA technology at that time, they weren't sure whether the scientific community would follow what the——

Mr. ABRAHAM. Much like now.

Dr. KAHN. Much like now. It's 40 years later and the scientific community is very different today than it was in 1975.

Mr. ABRAHAM. For sure.

Dr. KAHN. So we were undisputedly the leaders in the world of that science in 1975. As everybody has said, the technology now is much more widespread and much easier to implement, as Dr. Doudna has said, making it much more difficult for any one community of scientists to actually speak on behalf of the whole.

Mr. ABRAHAM. Well, we think—we can probably say realistically that America will lead in the discussion of the ethical and moral implications of this, but to follow up with your statement then, as Americans or as the United States——

Dr. KAHN. Um-hum.

Mr. ABRAHAM. —scientific community, if we should see an element get outside the bounds, can we do anything?

Dr. KAHN. Yeah, it's a great question. And as I said in my testimony and in my statement that you read and I reiterated today that the journal publication community has a very important role to play here. What—your research doesn't really mean much unless your peers have reviewed it, called it good, and then it gets published in a credible place. So that's a really important barrier. That doesn't mean people won't try to do things that aren't ethical and then try to get them published, but it's a—has a sort of strong inoculating feature I would say.

The other thing that Dr. Doudna pointed out but I'll reiterate is that there's been a little bit of a disagreement in the scientific community on the topic of gene editing, about whether there should be a moratorium only on clinical application or on something more widespread. And so that's a healthy debate to have and it's great that we're having it and it's great that the Academies are bringing it out to the international level. So that's exactly the discussion we ought to be having.

One last thing I'll say is that people around the world want to be part of the scientific community. There's a very strong incentive for them to behave, right, to follow the conventions which make them a legitimate member and I think we shouldn't underestimate the power of that. So, yeah, there might be fewer restrictions in other parts of the world but those people want to publish in Science and Nature just like American scientists do.

Mr. ABRAHAM. Yeah. Thank you.

Mr. Chairman, I'll yield back.

Mr. MOOLENAAR. Thank you.

I now recognize Mr. Sherman.

Mr. SHERMAN. Thank you, Mr. Chairman, for holding these hearings. This has been an area of intense interest on my part since the year 2000 when I went to the Floor and said that the most important decision we will make this century is whether our successor species is carbon-based or silicon-based, whether the new and intelligent species on this planet is the product of genetic engineering or the product of computer engineering.

Some of you will remember I served on this Committee in the 107th and 108th Congresses and this was pretty much my main reason for serving on the Science Committee.

I should bring to the attention of this Subcommittee that on June 19, 2008—transcript available—the relevant Subcommittee of Foreign Affairs had a hearing titled "Genetic and Other Human Modification Technologies: Sensible International Regulation or a New Kind of Arms Race?" And in fact the analogy to what we're talking about here is the only other technology that was equally explosive perhaps, and that is nuclear weapons technology. In 1939 Albert Einstein wrote to Roosevelt saying what was possible and policymakers had only six years before that technology literally exploded onto the scene. Thank God we've got a little bit more time but the Nonproliferation Treaty took many decades after 1939 or after 1945 and I think could be a good model for what we need here.

Dr. Kahn, it may be too long to list but I don't know whether America is exactly number one in this technology or whether Britain or China might be slightly ahead, but we're all within, I think, a few years, but there are a whole bunch of other countries either at that level or maybe four or five years behind. Can you even list the countries that within five years could be where we are now?

Dr. KAHN. I don't know—

Mr. SHERMAN. Are we talking a dozen, two dozen?

Dr. KAHN. I'm not sure that it's—that I or anybody could do that, and in fact, I'm not sure that it even requires countries. It's individuals who have access to the capacity—

Mr. SHERMAN. Um-hum.

Dr. KAHN. —which, as Dr. Doudna has said, is actually fairly democratic I think was the term that she used. So in a way, you know, it's about where people have the laboratory capacity and that's almost anywhere in the world.

Mr. SHERMAN. Well, thank God nuclear weapons take an industrial scale. And although they reflect only 1945 technology, we've seen in Iran that you have to do something big, you have to spend billions of dollars, it has to be visible to the world that you're doing something. You seem to be saying that what we're concerned about here could be a lot cheaper and take place in a laboratory basement?

Dr. KAHN. And I'll let my colleagues speak to the concrete answer to that question but I think exactly that kind of point is really important for how we think about what appropriate oversight, regulation, guidelines need to be—

Mr. SHERMAN. And I want to pick up on Mr. Foster's question about time frame, but one of the concerns we will have is that countries will see this as, oh my God, it might be terribly unethical but it gives us a leg up militarily or economically. Damn the torpedoes.

Leaving aside engineering intelligence and looking to things that would—other than that that would affect soldiers, soldier is a little better if they've got courage, stamina, and strength. We're already at a point where drugs are going to be used by various militaries to impart those characteristics to their soldiers but then we can go further to genetic engineering. What is the time frame before there's genetic engineering that would do the simplest of—I don't know which of those three is easiest—would give a soldier either more strength or more stamina or, say, more courage, more willingness to charge out? Is there any way to say that that soldier is five years from now, fifteen years from now, or are we in the world of science fiction? Anybody venture a guess?

Dr. McNALLY. Science fiction.

Mr. SHERMAN. Okay. Well, we'll be in a position I think next decade where at least—well, already many militaries are using drugs on their soldiers and then the next step will be to use the next element of medical technology, not drugs, but genetic engineering.

Does anyone have—I mean we're—the first and most ethical use of this technology is to remedy maladies. The next step will be to allow parents to have kids that have all the best characteristics of anyone in their family or in the world. And then we'll go to giving kids unprecedented capacities, and at that point we're talking maybe human, maybe trans-human. What—does anyone here have any view as to how long it will be before we can affect the intelligence of either people—either adults or of germ cell—the germ line?

Dr. McNALLY. I'll dive in.

Mr. SHERMAN. Dr. McNally is coming close.

Dr. McNALLY. I'll dive in.

As Dr. Doudna has said, the limitations of engineering things like intelligence—

Mr. SHERMAN. Um-hum.

Dr. McNALLY. —are far more limited by our genetic observations—

Mr. SHERMAN. Um-hum.

Dr. MCNALLY. —than they are by our capacity to do genome engineering. We have actually very large scale genetic projects ongoing right now, the 1,000 Genomes Project in the next year or two we'll be looking at a million human sequences and a lot of information connected to it. And the simple answer is that traits like intelligence are not single-gene—

Mr. SHERMAN. Yeah.

Dr. MCNALLY. They're probably not even entirely genetic is what people are referring to on a regular basis right now. There's a whole series of articles written about the missing heritability for many different traits, things that we thought were genetic when it turns out we look at the genes, they may not be so genetic or they are a very high complexity.

Mr. SHERMAN. So the technology—it's interesting because in nuclear weapons there's two components. One is weaponizing the highly enriched uranium, which turns out to be the easy part. It seems the most dangerous. Oh, my God, you're going to turn it into a nuclear weapon. The real hard part and the decider as to which countries have nuclear weapons technology is the ability to enrich uranium.

Mr. MOOLENAAR. The gentleman's time is—

Mr. SHERMAN. If I—it seems like what you're describing is a situation where the roadmap to what genes do—have what characteristics is the hard part and the part that will determine and the actual snipping and replacing and editing, you guys have that down.

I yield back.

Mr. MOOLENAAR. Thank you.

Ms. Bonamici.

Ms. BONAMICI. Thank you very much, Mr. Chairman. And thank you to the Chairman and Ranking Member for holding this hearing about this important topic, and I—my absence for most of the hearing was only because I was in another hearing. It does not indicate my lack of interest in the subject.

And I'm really glad that we have this, I agree, very incredibly qualified panel and I especially appreciate the gender diversity. As someone who works on education issues and trying to get more women in STEM, thank you for having a balanced panel.

So I missed—a lot of the questions have already been asked but there's a couple of things that I wanted to follow up on. There's been a lot of attention on using gene editing technologies in human embryos, and of course we—with the press from what happened in China, that's getting attention. But I know that a lot of the research is not in human embryos, so could you discuss how the technologies are being used in research today in other areas, including in organisms other than humans? And also can you talk about the potential promise from sectors other than healthcare, energy, for example?

Dr. DOUDNA. Okay. Well, I'll—I can take a stab at that. So you're absolutely right that the technology is being widely employed in many different kinds of cells and organisms, and I'll just give you a couple of examples. I think that in plant biology this is going to be, you know, equally impactful as the kind of thing that—kind of applications that we're talking about here in human health in terms of enabling very, you know, widespread introduction of genes into plants that could be beneficial especially for dealing with climate change and other kinds of environmental impacts in plants. And I'm just—I'm not a plant biologist but I'm saying this based on conversations I've had with people that are already doing those kinds of experiments and are extremely excited about the way that that research has been enabled.

And then you mentioned biofuels and I think that's very interesting because I'm aware of several groups that are actually using this kind of genome engineering for what we call systems biology, basically being able to make large-scale changes in the genomes of organisms that will be useful for producing various kinds of chemicals, including, you know, biofuels and other very important materials that can be difficult to obtain in other ways.

Ms. BONAMICI. Terrific. And I know Dr. Dzau wants to——

Dr. DZAU. This in fact is a very important area. Today we're talking about human genome editing but I'm glad you raised the question about the usage in other living organisms. Mr. Sherman asked about, you know, what will be the misuse, if you will, but we do have to think about, you know, the use in plants, insects, et cetera, you know, what happens if there's misuse and what would happen to the environmental impact and who actually gets to decide who's going to do what? And there are more commercial sources I believe—opportunity for commercialization in changing coffee beans or whatever that you can imagine.

So this is an area that in fact—called gene drive that, again, National Academies are looking very carefully at this and how to regulate this. So I do think this is to me a very important area even though our focus is on human gene editing.

Ms. BONAMICI. Terrific. And I'm going to slightly change the topic.

Dr. Doudna—did I say your name properly?

Dr. DOUDNA. Yes.

Ms. BONAMICI. I know you cofounded several startup companies and we've had conversations in this Committee before about the challenges in trying to launch companies but also transferring academic research into the marketplace. So can you chat a little bit about that and what the challenges have been and what the Federal Government can do to help with transferring academic research into the marketplace?

Dr. DOUDNA. Yeah, this is a very important area and I'm a real newbie to it. This is actually the first time that research in my lab has led to something that, you know, had clear commercial applications. So I can tell you my experience, you know, being involved in starting companies and raising money for companies. We've had different experiences I would say. I've been involved in three different startups around this technology so far. And I've had a lot of help with one of them in particular through a—what would I call it? I guess it's a biomedical research initiative in the Bay area that was actually funded in part by the State of California. And what that does is to give people like me who know nothing about, you know, starting a business some training. We got access to some legal advice early on. We—I think this Institute paid for the incorporation of the company initially and then gave us some support in terms of writing for federal funding for the company.

I think this kind of support is really important. And as I talk to my colleagues around the country who are trying to do similar things, commercializing work coming out of their academic labs, I hear over and over how all of us who are in the academic world could really benefit from that kind of level of support. I'm not sure if it's something that happens at the federal level or if it's better done lower down, hard to say.

Ms. BONAMICI. Thank you so much.

And my time is expired. I yield back, Mr. Chairman. Thank you.

Mr. MOOLENAAR. Thank you.

And now I want to recognize Mr. Palmer.

Mr. PALMER. A couple of quick questions. Most of this discussion I think has been about government research. You're talking about convening an international summit. I just want to make sure that when we're talking about implementing safeguards that this includes the private sector, that they're not running out there as mavericks. Is that correct?

Dr. DZAU. Certainly our intention, if you look carefully at the way we put our advisory committee together, we have people who have great industry experience on it and our intention is to include industry and, you know, people in the commercial sector to be in this discussion.

Mr. PALMER. Thank you. I yield the balance of my time, Mr. Chairman.

Mr. MOOLENAAR. Thank you.

I now recognize Mr. Foster.

Mr. FOSTER. Thank you.

It's my understanding that very often there's a computing bottleneck in our ability to analyze genomes. And I guess, Dr. McNally, can you tell the Committee about some of the work you've done at Argonne——

Dr. MCNALLY. Yeah.

Mr. FOSTER. ——which is a facility shared by the Ranking Member and myself.

Dr. MCNALLY. Yeah, so right now it's—since more than a year-and-a-half, it's been very possible to sequence a human genome with the consumable cost being about $1,000, which is an amazingly low price for a single genome. That doesn't include the cost of analysis, and so the real rate-limiting step right now is the time it takes to analyze genomes.

So what we've done is we've actually, working together with Argonne National Labs, the University of Chicago, and now Northwestern, we've actually taken a Cray XE6 supercomputer that's at Argonne and outfitted it with all the computing code that's available through the Broad Institute so that we can now, for example, analyze 250 genomes in a weekend if they give us the whole machine to work. So that dramatically accelerates our ability to screen through three billion bases of an individual genome and score it's four million differences that exist in each one of those genomes, the vast majority of which are rare and private.

Mr. FOSTER. And—well, thank you. That's very interesting. I guess another application of federal investments that are having an unintended benefit.

I guess for Dr. Kahn, has anyone ever taken the bull by the horns and actually attempted to draft legislation or international treaties, you know, at regulating human genetic engineering or other genetic engineering in the environment?

Dr. KAHN. That's a good question. I don't know that there have been legislation or, you know, bills proposed and I don't know about international treaties. There certainly have been efforts to craft guidelines but they're nonbinding by entities like the World Health Organization or the World Medical Association, which has crafted actually fairly well-known guidelines on human subjects research which is called Declaration of Helsinki, which tends to be followed but in a voluntary way and then built into regulation at the national level. But I don't think that there has been any successful international regulation.

Mr. FOSTER. All right, or even proposed—just—you know, lists enumeration of all of the issues that you have to resolve when you write rules.

Dr. KAHN. Yeah, I'm not aware, although not a historian of science. That's a really interesting question that I—maybe I'll do some looking and see if I can get back to you.

Mr. FOSTER. I would appreciate that.

Dr. KAHN. Sure.

Mr. FOSTER. And do you understand right now if you're on a hospital ship in international waters, who's the regulator?

Dr. KAHN. That's the right kind of question to be asking. And this sort of issue came up when there was a supposed cloning of a human being. You may remember back in—when that—whenever that was, the late '90s, early 2000s—

Mr. FOSTER. All the——

Dr. KAHN. Yeah, which turned out to be a hoax. But one of the claims was it was being done in international waters and out of the reach of any of the national regulations or governance structure, which, you know, in a way was a helpful aha moment, right? You need to think about what to do when all you need is a well-fitted ship that has the laboratory on board and you're outside of the—any restrictions that an international treaty or national government might employ.

Mr. FOSTER. Okay. And I guess one last question. The issue of gene drives has entered the news and interestingly is with the potential to sort of take over the genome of an entire species in the wild in the course of, you know, a few dozens of generations. And so this obviously has huge environmental effects and has to be internationally regulated presumably because of, you know, insects don't normally, you know, obey national borders.

And so I was just wondering if that is an area where, for example, this Committee might interestingly have a separate branch of investigation that all of the applications of this technology to plants and animals in the wild and in the laboratory?

Yes, Dr. Dzau.

Dr. DZAU. Mr. Foster, as I mentioned, I think the National Academies is undertaking such a study. We'd be happy to send you all the materials, and in fact, as we look at what we are potentially covering and not covering, we can have a conversation about what else should be done.

Mr. FOSTER. Okay. And so the charge is complete for the study or is it——

Dr. DZAU. Yes.

Mr. FOSTER. ——ongoing? All right. I'd appreciate that. Thank you. I yield back.

Mr. MOOLENAAR. Thank you.

And I'd like to thank our witnesses for the testimony today, outstanding, and all the Members for the questions. The record will remain open for two weeks for additional comments and written questions from Members. The witnesses are excused and this hearing is adjourned.

[Whereupon, at 3:56 p.m., the Subcommittee was adjourned.]

Appendix I

ANSWERS TO POST-HEARING QUESTIONS

Answers to Post-Hearing Questions

Responses by Dr. Elizabeth McNally

HOUSE COMMITTEE ON SCIENCE, SPACE, AND TECHNOLOGY
SUBCOMMITTEE ON RESEARCH AND TECHNOLOGY

"U.S. Surface Transportation: Technology Driving the Future"

Questions for the record, Dr. Elizabeth McNally, Professor of Genetic Medicine, Professor in Medicine-Cardiology and Biochemistry and Molecular Genetics; Director, Center for Genetic Medicine, Northwestern University

Questions submitted by Rep. Elizabeth Esty, Member, Research and Technology Subcommittee

1. You work with couples who are facing extremely difficult decisions about starting or growing their families. Knowing how far we have advanced with genetic editing, and the possibilities you outlined with pre-implantation genetic diagnosis, how do you think physicians should balance their patients' desire to protect their children with ethical concerns surrounding this technology?

Response: Physicians should counsel patients and families by weighing multiple issues. Technology and its medical application are judged based on the risk-benefit ratio. At this point in time, the technology for human genetic engineering is not sufficiently advanced, so risks including off-target effects and inadequate efficiency outweigh benefit. However, we can anticipate that this risk-benefit ratio will shift in the future towards less risk and more benefit.

With technology advances, the risks will include cost. In the case of preimplantation genetic correction, the cost will likely be borne directly by families rather than insurers. Additional risks include ethical concern and indeed this was the major topic of the hearing. It is likely that the physicians who will discuss these choices with families will be those highly expert in genetics. For example, preimplantation genetic diagnosis is usually provided in highly specialized centers where additional counseling, information, and guidance are provided to families to aid in their decision-making. The ethical aspects associated with these choices are often discussed by genetic counselors.

Some risk comes from physicians who may not be sufficiently educated on the technology. The lack of genetic knowledge among physicians is growing to be an increasing problem simply because the field and technology are moving quickly. The American College of Medical Genetics and National Society of Genetic Counselors are aware of this growing need, and it is generally recognized that more education will be needed for physicians and non-physician providers moving forward.

Appendix II

ADDITIONAL MATERIAL FOR THE RECORD

STATEMENT SUBMITTED BY FULL COMMITTEE RANKING MEMBER
EDDIE BERNICE JOHNSON

Thank you, Madam Chairwoman for holding this hearing, and I want to join you in welcoming our distinguished panel of witnesses.

This afternoon we are talking about new gene editing technologies that have promising applications in fields ranging from medicine to energy to agriculture. The Chinese research paper that prompted this hearing highlights the need to have a serious examination of the science, safety, and ethics of gene editing technologies.

I want to thank the Chairwoman and Ranking Member for putting together this hearing with an impressive panel of expert witnesses. Additionally, I want to acknowledge Dr. Foster, whom I understand was instrumental in advocating for this hearing.

The technologies we are discussing today can alter DNA—the blueprint of life. There are significant safety, efficacy, and ethical issues concerning these technologies. What are the ethical uses of these technologies? Is it ethical to use them if they could cure a disease? What if they just treated a disease?

Any applications that would alter germline cells, where changes are passed down through generations, have additional ethical issues. For example, is it ethical for the current generation to consent to changes for a future one?

Although today we are discussing human applications, we should not forget that these same technologies also have great potential for use in energy and agriculture. Gene editing could be used to create biofuels and new crops. Responsible applications of these technologies could lead to significant economic growth if the U.S. takes the lead in research and transferring that research to the private sector.

I look forward to hearing from the witnesses about the state of the gene editing science and potential applications. Additionally, the U.S. needs to take a leadership role in addressing relevant ethical issues so I am glad that the panel includes a bioethicist to help us better understand what is involved.

Finally, I look forward to hearing about the National Academies' plans to address these and other important questions surrounding this emerging research area.

Thank you and I yield the balance of my time.

www.ingramcontent.com/pod-product-compliance
Lightning Source LLC
Chambersburg PA
CBHW080538190526
45169CB00007B/2540